T0094239

Fixed Offshore Platforms

Structural Design for Fire Resistance

Fixed Offshore Platforms

Structural Design for Fire Resistance

Mavis Sika Okyere

CRC Press
Taylor & Francis Group
Boca Raton London New York

CRC Press is an imprint of the
Taylor & Francis Group, an **informa** business

CRC Press
Taylor & Francis Group
6000 Broken Sound Parkway NW, Suite 300
Boca Raton, FL 33487-2742

© 2018 by Taylor & Francis Group, LLC
CRC Press is an imprint of Taylor & Francis Group, an Informa business

No claim to original U.S. Government works

Printed on acid-free paper

International Standard Book Number-13: 978-1-138-60133-8 (Hardback)

Visit the Taylor & Francis Web site at
http://www.taylorandfrancis.com

and the CRC Press Web site at
http://www.crcpress.com

Contents

Preface

The purpose of this book is to present the methodology that will enable an engineer to structurally design a fixed offshore platform to resist fire. It is intended to highlight the principal considerations of design and to establish a common method in carrying out the design.

Today, computers provide engineers with powerful tools for analysing large and complex structures with greater precision and speed than was ever possible in the past. However, the results of these analyses will obviously be incorrect if an error is made in the input data or if the structure is not properly modelled. Thus, to avoid costly failures or expensive repairs to structures that prove to be incorrectly designed, engineers must develop an intuitive sense of correctness. A goal of this research is to help readers develop an intuitive understanding of structural analysis of a fixed offshore platform and recognize when a solution is or is not accurate.

Author

Mavis Sika Okyere (neé Nyarko) is a pipeline integrity engineer at Ghana National Gas Company. She holds a bachelor of civil engineering degree from Kwame Nkrumah University of Science and Technology and a master of science in gas engineering and management from the University of Salford. Okyere has worked with LUDA Development Ltd, Bluecrest College, INTECSEA/Worleyparsons Atlantic Ltd, Technip, Ussuya Ghana Ltd and the Ghana Highway Authority.

Her main research areas include pipeline flow analysis (specifically, gas flow analysis, hydrocarbon liquid flow analysis and pipeline flow improvement), engineering design of oil and gas pipelines, and the sizing of gas pipelines and the stress-based design of pipelines.

Introduction

1.1 GENERAL

In the event of a major fire incident, it is our principal objective to be able to maintain the fixed offshore platforms integrity for an adequate period of time to allow fire and damage-control measures to stop continued deterioration of structures, while allowing evacuation of operation and maintenance personnel. This involves the analysis of policies or strategies and alternative methods within the context of defined restraints and service requirements.

Protecting people from injuries is linked to designing structures. Structures should be designed to withstand loads without creating dangerous fragments or falling down during any fire incident. The fixed offshore platform must be able to sustain its safety functions for a required performance time if subjected to fire and explosion damage. Therefore, there is a need to design the offshore structure to have redundant elements appropriately placed, be able to withstand increased loadings and have the ability to redistribute loadings.

In addition to designing the platform to resist fire, safety engineers should use active and passive fire protection measures to protect the structure. Escape routes and safe areas should be maintained to allow sufficient time for platform evacuation and emergency response procedures to be implemented.

The main cause of fatalities in fire is smoke, and most deaths occur long before there is any significant risk of structural collapse. Therefore, automatic fire sprinklers can be installed on the platform to automatically suppress small fires on, or soon after, ignition/explosion or to contain fires until the arrival of the fire service.

Structural design for fire resistance involves

- Considering fire and explosion as a structural load.
- Assessing the structural fire resistance.
- Fire, explosion and blast effect analyses. Estimate the response of the structures subjected to the load from hydrocarbon explosion.

- Designing the offshore structure to have sufficient residual strength and structural robustness to meet the acceptable level of risk.
- The slenderness ratio and the wall thickness modulus.
- Fire design of structural connections.
- Enforcing safety on the platform.

The slenderness ratio and the wall thickness modulus are used by offshore structural engineers with regard to structural fire endurance. Fire, explosion and blast effect analyses should be performed together.

For existing sites, active or passive mitigation techniques may be applied. They include those which could be activated following a release of flammable material in order to prevent flame acceleration, such as water curtains. Alternatively, pipe racks could be artificially divided into regions too small for flame acceleration sufficient to allow high speed flames to occur (Harris 1983). The design against accidental fire situations should be included in the structural design of the fixed platform in collaboration with safety engineers.

1.2 OBJECTIVE

The main objective is to reduce to within standard limits the possibility for death or injury to the occupants of a fixed offshore platform and others who may become involved, such as maintenance and operation personnel and the fire and rescue team, as well as to protect contents and ensure that the platform can continue to function after a fire and that the platform can be repaired.

1.3 FIXED OFFSHORE PLATFORMS

A fixed offshore platform is a type of offshore platform used for the production of oil or gas in shallow waters. It extends above the water surface and is supported at the seabed by means of piling or a shallow foundation, with the intended purpose of remaining stationary over an extended period. Fixed offshore platforms are anchored to the ocean floor and provide stable working platforms for the production of oil and natural gas.

A fixed platform may be made up of three parts: foundation, a welded tubular steel jacket and a deck or topside (including any surface facility). Piles are driven into the seafloor to secure the jacket. The water depth at the intended location dictates the height of the platform. After safeguarding the jacket and installing the deck, supplementary modules may be added for drilling, production and crew operations.

Jackets consist of tubular members that are welded together with pipe braces to form a stool-like structure. The jacket is safeguarded to the seafloor by weight and a piled foundation that penetrate hundreds of metres beneath the mud line. Skirts may also be added to help in fixing the jacket to the seafloor. A pipeline is connected to the jacket through a jumper at its base. The pipe connects the platform to shore or another facility.

Fixed platforms have an advantage of stability because they are attached to the seafloor, and there is limited exposure to movement due to wind, current and wave forces.

Financial considerations limit construction of fixed platforms to water depths not greater than 2000 ft (610 m) because it is simply not economical to build legs that are too long.

1.3.1 Structural Components of a Fixed Offshore Platform

The three main structural components of a steel jacket offshore platform are the deck, jacket and foundation (Figure 1.1).

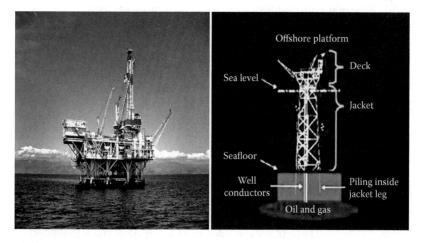

FIGURE 1.1 Different parts of a fixed offshore platform with a pile foundation.

1.3.1.1 Deck

The main structural components of a deck are

- Deck beams
- Deck legs
- Deck plate and or grating
- Skid beams (if drilling using a tender rig is planned)
- Main (longitudinal) and wind trusses (if the deck structure is to be an open girder system, this function will be performed by a portal frame)

The deck provides support for the drilling, production equipment and life-support system of the platform. Due to operational requirements, fabrication infrastructure and installation equipment availability, various types of decks have been developed. An integrated deck is designed as a single-piece structure, while a module support frame (MSF) supports few large modules. MSFs can be built in areas with readily available fabrication infrastructure and equipment.

An integrated deck may be divided into a number of levels and areas depending on the functions they support (Corr and Tam 1998). Typical levels are

a. Main (upper) deck: It supports the drilling/production systems and several modules (drilling, process, utilities, living quarters, compression, etc.).

b. Cellar deck: It supports systems that need to be placed at a lower elevation and installed with the deck structures, such as pumps, some utilities, pig launchers/receivers, Christmas trees, wellhead manifolds, piping, etc.

c. Additional deck levels, if needed. For example, if simultaneous drilling and production operations are planned, some process equipment may be located in a mezzanine deck.

A modular deck may be classified as

a. Module support frame (MSF): Provides a space frame for supporting the modules and transferring their load to the jacket/tower structure. MSF may also be designed to envelope a number of platform facilities, such as the storage tanks, pig launching and receiving systems, metering/proving devices and the associated piping systems.

b. Modules: These provide a number of production and life-support systems, such as

i. Living quarters: Support a heliport, communication systems, hotel messing, office and recreational facilities.
ii. Wellhead module: Supports the wellheads, well test and control equipment.
iii. Drill rig module: Contains the drill tower, draw-works, drillers and control rooms, drill pipe and casing storage decks and pipe handling systems. It is located over and supported by the wellhead module.
iv. Production module: Contains oil/gas/water separation and treatment systems and other piping and control systems and valves for safe production, metering and transfer of the produced liquids and gas to the offloading system.
v. A compression module: It may be added if gas compression is needed.

1.3.1.2 Jacket

A jacket provides support for the deck, conductors and other substructures such as boat landings, barge bumpers, risers, sumps, walkways, J-tubes, mudmats, and so on. The main structural members of a jacket are

* Braces
* Joints
* Jacket legs
* Launch runners and trusses
* Skirt pile sleeves and braces (only required if skirt piles are needed)
* Appurtenances such as conductor bracing and guides, risers, clamps, grout and flooding lines, J-tubes, walkways, mudmats, barge bumpers, boat landings

1.3.1.3 Foundation

Detailed design of the foundation is described in subsequent chapters.

1.3.2 Types of Fixed Offshore Platforms

1.3.2.1 Template or Jacket Platforms

Jacket platforms consist mainly of a jacket, decks and piles. All of the petroleum platforms installed in the Persian Gulf are the template (jacket) type. It is a space-framed structure with tubular members supported on piled foundations.

1.3.2.2 Tower

A tower platform is a modification of a jacket platform and may or may not be supported by piling. Piling may serve as well conductors and may be driven through sleeves inside or attached to the outside of the leg.

1.3.2.3 Gravity Platforms (Concrete Gravity Structures)

Gravity platforms are fixed-bottom structures which are made of concrete. These platforms are so heavy that they do not have to be physically attached to the seafloor (no need for piled foundation), but instead, simply rest on their own mass to resist environmental loads.

There are more than 26 concrete gravity-support structures for oil- and gas-producing platforms in the North Sea. These are used at a moderate water depth of 40–303 m and require less maintenance. The design of the concrete gravity structures inevitably involves the use of finite element (FE) analysis. Before establishing the FE model, the geometry must be fixed and all major loads should have been determined.

TABLE 1.1   A summary of major fire incidents on offshore platforms

DATE	PLATFORM	FATALITIES	TYPE OF INCIDENT
1969	Elefante	0	Fire
1970	Main Pass Block 41	–	Fire
1975	Ekofisk A	6	Explosion/fire
1983	Nowruz Platform	20	Fire
1984	Petrobras Enchova Central	37	Blowout/explosion/fire
1984	Getty Platform A	1	Explosion
1987	Steelhead Platform	0	Blowout/fire
1988	Petrobras Enchova Central	0	Blowout/fire
1988	Piper Alpha	167	Explosion/fire
1989	Cormorant Alpha	–	Explosion
1989	Ekofisk P	0	Fire
1991	Fulmar A	3	Explosion
1996	Sundowner 15	–	Fire
1996	Ubit Platform	18	Explosion/fire
1997	Pride 1001E	–	Blowout/fire
1999	NFX Platform A	–	Blowout/Fire
2001	Petrobras P7	0	Blowout/fire
2001	Petrobras P36	11	Explosion
2005	Mumbai High North	22	Fire

FIGURE 1.2 Piper Alpha Oil Rig at its end. (Image reused with permission from Raabe, D. n.d. Dramatic failure of materials in drilling platforms. Retrieved August 12, 2017, from http://www.dierk-raabe.com/dramatic-material-failure/oil-drilling-platforms/)

1.4 MAJOR FIRE INCIDENTS ON OFFSHORE PLATFORMS

Learning from past disasters will help us know the measures to undertake to prevent future catastrophes. Fires and explosions that occur on oil and gas platforms are some of the most disturbing types of offshore accidents that can ever occur. These fires and explosion incidents can result in extreme injuries and property damage and can result in long-term consequences. Some of the largest oil and gas platform fires and explosions that have occurred in the twentieth and twenty-first centuries are shown above (Arnold 2017), (Table 1.1, Figure 1.2).

Design
Methodology

2

The structural design of an offshore platform falls into the following four phases/stages (SJR 1994):

    a. Input data
    b. Data collection and conceptual/preliminary design
    c. Detailed design
    d. Design output

2.1 INPUT DATA

The following input data may be required for the design of an offshore platform:

    a. Functional requirements
    b. Design constraints
    c. Environmental data

Examples of functional requirements are

- Number and spacing of wells
- Number of risers
- Number of pipelines
- Piping and manifold requirements
- Maintenance and service
- Levelling and positional requirements
- Protection requirements
- Operational
- Development program
- Corrosion

Examples of design constraints are

- Member sizes selection
- Steel section selection (e.g. grade of material, section type, etc.)
- Fabrication yard restrictions (e.g. tolerances, fabrication site, etc.)
- Installation and transportation equipment (e.g. vessels, cranes, etc.) and their availability
- Codes, standards, certifying authorities and client requirement

Examples of environmental data are

- Metrological
- Soil data
  - Soil friction angle
  - Soil density
  - Soil shear strength
  - Soil friction (lateral and longitudinal)
- Seabed topography
- Marine growth
- Bathometry
- Wave height
- Current
- Wind
  - Wind velocity (m/s) with return periods (years)
  - Wind direction (from platform north)
  - Ambient air temperature (degrees Celsius, minimum and maximum value)
- Seismic activity
- Erosion
- Seawater
  - Water density (kg/m³)
  - Water kinematic viscosity (m²/s)
  - Marine growth elevations (m), thickness (mm) and density (kg/m³)
  - Seawater temperature (degrees Celsius)
  - Water depth
    Water depth and variations (m) with return periods (years)—
    LAT (lowest astronomical tide) water depth, HAT (highest astronomical tide) above LAT, the extreme water depth, storm surge, design still water level above LAT, design crest elevation. When specifying the water depth for a fixed offshore platform, account should be made for any settlement or subsidence

- Ice (fixed, floes, icebergs)
- Earthquakes (if necessary)

2.2 DATA COLLECTION AND CONCEPTUAL/PRELIMINARY DESIGN

This stage of the project is also known as front-end engineering design (FEED). The basis of design is prepared at this stage of the project. This phase of the design process requires the collation or gathering of the data required for establishing functional layout and member sizes and for the detailed design (SJR 1994). Examples of this data include

- Equipment weights
- Equipment layout (preliminary equipment sizing and area requirement)
- Access and clearance criteria
- Environmental loading data
- Operational loads
- Accidental loads
  - Fire and blast load
- Lift and installation methods and loads
- Soil and foundation data
- Design constraints
- Design philosophy
- Corrosion
- Estimation of CAPEX (capital expenditure) within ±40%

The structure must be capable of safely withstanding some or all of the following load conditions (SJR 1994):

- Lifting and lowering in the air and water
- Transportation over land and/or sea
- Temporary and permanent support on the seabed
- Pilings
- Equipment installation
- Operating loads
  - Equipment and operating loads
  - Environmental loads (waves, currents, wind)

- Accidental damage
- Foundation settlement

2.2.1 Service Requirements

The layout of the structure must allow for ease of

- Equipment installation
- Equipment removal
- Maintenance
- Diver and ROV (remotely operated vehicle) accessibility
- Diver and ROV working space
- Tying into other installations

The design of the structure must consider the cost implications concerning the choice of

- Layout
- Materials
- Fabrication
- Lift and loadout
- Transportation
- Installation

The life of the structure must be considered with regard to

- The required design life
- Accident damage
- Corrosion
- Fatigue

The conceptual design may commence after establishing basic loadings and layout requirements. The preliminary design should be based on the most critical load cases only.

Wherever possible, reference should be made to past project experience and details of existing similar work. The content and presentation of a preliminary design should be simple and slightly conservative for the following reasons:

- Speed and economy of design
- Some design data may not be available until late into the design process. Examples include
  - Soil data
  - Transportation data

- Well spacing
- Environmental data
- A number of preliminary designs may be required in order to arrive at an acceptable design
- Some data may be preliminary or subject to alteration. Examples include
  - Additional equipment loads
  - Alteration of development program

The selected design should

- Satisfy the functional requirement
- Be sized to cater for lifting, installation and in-service conditions
- Cater for the foundation requirements for in-service and installation conditions
- Contain a weight take-off or weight estimate

In addition to the selected design, the following should be established to form the basis of the detailed/final design:

- Design database
- Codes, standards and certification requirements; the major certification bodies are
  - Det norske Veritas (DnV)
  - Lloyd's Register of Shipping (LRS)
  - American Bureau of Shipping (ABS)
  - Bureau Veritas (BV)
  - Germanischer Lloyd (GL)
- Design methods
- Design requirements and philosophy

2.3 DETAILED DESIGN

The detailed design involves the following:

a. Primary structure design
   Analysis of the overall structure for the following load cases:
   - In service
   - Lifting in air
   - Lowering in water

- Transportation
- Installation
- In service
- Impact/accidental damage

A computer-based FE analysis software such as SACS may be used to perform the design of the primary structure.

b. Local design of
- Joints
- Lifting attachment
- Pull-in points
- Tie- and pull-in points
- Local impact
- Equipment supports
- Levelling supports
- Local buckling
- Hydrostatic collapse
- Member fatigue

The local design will normally be done by hand calculation.

c. Structural fire-endurance assessment
- Fire-resistance assessment
- Heat transfer analysis
- Structural (mechanical fire) analysis
- Robustness
- Residual strength
- Fire-resistance test
- Empirical correlation
- Fire, explosion and blast effect analysis

d. Preparation of drawings

e. Detailing of reinforcement and reinforcement drawings

f. Installation
- Lifting arrangements
- Levelling and positioning
- Tolerances
- Flooding design
- Pile driving
- Pile fixing

- Weight control and determination of centre of gravity (and centre of buoyancy where necessary)
- Corrosion protection

2.4 DESIGN OUTPUT

The design output may consist of all or part of the following:

- Design report (including computer output)
- Drawings
- Installation specification
- Fabrication specification
- Material specification
- Levelling method
- Instrumentation specification
- Material take-off
- Anticorrosion specification
- Fire-resistance specification
- Specification for other safety components

Design Requirements

3

This chapter should be read chronologically, as it is important to establish all the necessary data and functional requirements before analysing the structure.

3.1 DESIGN CODES

The platform design, analysis, fabrication, installation, use, maintenance, inspection and assessment must conform to

- ACI Standard 318-95: Building Code Requirements for Reinforced Concrete (ACI 318-95) and Commentary (ACI 318R-95)
- API Recommended Practice (RP) 2A-WSD: RP for Planning, Designing, and Constructing Fixed Offshore Platforms – Working Stress Design
- NACE Standard RP 0176-2003: Standard RP for Corrosion Control of Steel Fixed Offshore Platforms Associated with Petroleum Production
- ACI 357R-84: Guide for the Design and Construction of Fixed Offshore Concrete Structures, 1984 (reapproved 1997)
- ANSI/AISC 360-05: Specification for Structural Steel Buildings
- API Bulletin 2INT-DG: Interim Guidance for Design of Offshore Structures for Hurricane Conditions
- API Bulletin 2INT-EX: Interim Guidance for Assessment of Existing Offshore Structures for Hurricane Conditions
- API Bulletin 2INT-MET: Interim Guidance on Hurricane Conditions in the Gulf of Mexico
- ASTM Standard C 33-07: Standard Specification for Concrete Aggregates (approved December 15, 2007)
- ASTM Standard C 94/C 94M-07: Standard Specification for Ready-Mixed Concrete (approved January 1, 2007)

- ASTM Standard C 150-07: Standard Specification for Portland Cement (approved May 1, 2007)
- ASTM Standard C 330-05: Standard Specification for Lightweight Aggregates for Structural Concrete (approved December 15, 2005)
- ASTM Standard C 595-08: Standard Specification for Blended Hydraulic Cements (approved January 1, 2008)
- AWS D1.1: Structural Welding Code – Steel
- AWS D1.4: Structural Welding Code – Reinforcing Steel
- AWS D3.6M: Specification for Underwater Welding
- NORSOK Standard N-004: Design of Steel Structures (2004)
- Army Technical Manual TM5-1300: Structures to Resist the Effects of Accidental Explosions (1990)
- API RP 2FB: Recommended Practice for the Design of Offshore Facilities against Fire and Blast Loading (2006)
- ISO 19902: Petroleum and Natural Gas Industries – Fixed Steel Offshore Structures (2007)

3.2 FUNCTIONAL REQUIREMENTS

3.2.1 Drilling and Production Templates

Drilling templates are required for the positioning of wells. They are also used for supporting wellhead receptacles and as a docking reference when installing a jacket over a wellhead. For the purpose of docking, a jacket or docking piles may be fixed relative to the drilling template.

Production templates are required to house and support a combination of drilling, production and control equipment.

The offshore platform is used mainly in drilling for producing offshore oil and gas. The template must be capable of accurately positioning and supporting the equipment associated with the drilling and production of the wells. The template structure is normally specified by its length, width, height, weight and well spacing.

Fixed platforms may have as few as three or more than 50 well conductors. Usually, the drilling rig is not a permanent part of the fixed structure. But, on some occasions, the unit is left on the platform for future workovers or additional drilling if removing it is uneconomical.

Well spacing may vary according to operational circumstances from 2.5 m to about 7.5 m.

3.2.1.1 Levelling

If the structure needs to be levelled, mudmats and hydraulic jacks are usually used in conjunction with ROV- or diver-operated control panels. However, pile jacking systems are becoming more common.

If possible, a three-mudmat support system is preferable, arranged such that the distribution of load is approximately equal. A variation on this is to use four mudmats and to hydraulically link two adjacent hydraulic jacks to effectively create a three-point support system (SJR 1994).

It is important to remember that the whole jacking/levelling system requires testing and commissioning.

3.2.1.2 Location and Orientation

The subsea facility will have to be located on the seabed within certain tolerances of

- Level, to ensure verticality of wells
- The required geographical coordinates, to meet drilling and reservoirs engineering constraints
- The required orientation, to meet flow and control line tie-in restraints and drilling criteria

The orientation may also be selected to minimise environmental loads on the subsea facility and the floating production or service vessel, and to minimise deposition of surface returns of drilling cuttings on the subsea facility. This information is likely to be provided by the template designer.

The accuracy required for the positioning of the template could govern the method of handling and controlling the template during lifting, lowering and placement. It will also govern the instrumentation which will be required to monitor and measure the placement of the template.

3.2.2 Protection Structure

A protection structure may either be incorporated into a template or support structure or may be a separate structure.

If it is separate, it can either be supported directly on the seabed or on mudmat or piles, or it can be landed onto the support structure.

The most important requirements of a protection structure are that it should not interfere with the operation of whatever is being protected and that it should provide protection from the anticipated accident scenarios.

The structure may be required to be retrieved to the surface to allow access to the equipment being protected, or alternatively, the structure may incorporate lids. The lifting load of either of these options should preferably be kept within the limits of a diving support vessel or similar offshore support vessels (SJR 1994). If the protection structure is to be loaded onto the support structure, docking will normally be achieved by guideposts and cones. If the structure is symmetrical, the protection cover should be capable of docking in either orientation. If the structures (skid and cover) are to be piled, consideration may be given to making at least one pile common between the two structures, thus ensuring that the two are linked and preventing possible large relative movement (which can cause clearance problems between the protection cover and skid piping components).

3.2.2.1 Clearance/Access

The protection structure should allow clear overhead access to valves and other major equipment (with covers removed) such as

- Workover frames (in addition, allow minimum 0.3 m all round for installation)
- Maintenance envelopes (diver or ROV)
- Hydraulic accumulators and/or junction boxes
- Stab-in plates and cones
- Valve actuator, bonnet, valve internals or cover removal and replacement
- Valve tooling receptacle access (ROV)
- Hyperbolic welding habitats (spool and valve tie-ins)

Side access may also be required for

- Valve actuator umbilical
- Tooling interfaces
- Diver/ROV access
- Stab-in plates and cones

Actual clearance and access requirements should be obtained from diving consultants and/or valve and equipment manufacturers, as necessary.

3.3 PROTECTION

Protection planning includes structures to protect the offshore platform against dropped objects, such as anchor chains and trawl gear, or any other damages during installation in restricted areas with regard to trawling. The protection structure may also be designed as fully overtrawlable (NORSOK Z-014). The structural layout is to be planned with due consideration to the possibility of fire, explosion and other accidental damage.

Pipes and equipment, damage to which involves risk of explosion, fire or extensive pollution, are to be protected so as to minimise the risk of accidental damage. The protection may be established by providing a sheltered location, by local strengthening of the structure or by appropriate fender systems. Risers and similar equipment are to be positioned at a safe distance from boat landings, from cranes where objects may be dropped by accident, and so forth.

The following can provide protection against accidental damage:

3.3.1 Riser Guards

Riser guards consist of tubular steel space frames installed on offshore platforms to provide protection to risers against vessel collision or any accidents that may occur. As risers transport the crude oil which is highly flammable, a collision between vessel and risers has the potential of causing severe structural damage to the platform and injuries to on-board service personnel.

Conventional riser guards are designed to withstand impact energy from accidental vessel collision of up to 2.25 MJ.

A static in-place analysis of the riser guard using a structural computer program should be performed. Members should be analysed in accordance with the requirement of AISC.

3.3.2 Overtrawlable Structures

Normally, a protection structure is included to protect the equipment against dropped objects. The protection structure may be designed to be overtrawlable; that is, it protects the trawl gear against damage.

Overtrawlable structures are used to protect wellheads, Christmas trees, manifolds and other subsea assets from harsh environments and maritime activity and should be designed according to the requirements of NORSOK Standard U-001/ISO 13628-1.

The overtrawlable type of structure has a profile of a truncated pyramid (i.e. a looped 'A' shape).

For overtrawlable structures, the following requirements of NORSOK U-002 shall apply:

a. The protective structure shall deflect all fishing equipment.

b. The structure shall include corners, with the maximum true angle of 58° from the horizontal optimised to assist trawl and trawl wire deflection.

c. Corners, ramps and equivalent structures shall penetrate the seabed to avoid snagging from trawl warp lines and ground rope. Effects from installation tolerances and expected scouring shall be accommodated.

d. The overall geometry of the structure and the size of openings shall be such that trawl doors are prevented from entering into the structure.

e. If vertical side bracings are included, these shall be spaced to prevent intrusion and rotation of trawl equipment, without restricting subsea structure access for the intervention systems.

All protuberances shall be minimised to prevent snagging of nets. The lower the structure, the less effect the trawl gear will have on friction, pullover and snagging. All external edges/members shall have a minimum radius of 250 mm.

3.3.3 Ice Barriers

For the protection of offshore production facilities in shallow water areas, permanent ice barriers made from rock or concrete structures are often the appropriate solution.

For drilling exploration facilities, which usually have to change their location after a well is sunk, ice barriers should be made of steel. Steel ice barriers can easily be installed and de-installed.

The ice protection structure must be robust and simple to withstand hard environmental conditions and large ice loads.

3.3.4 Lateral Impact Protection

Protection against lateral impact due to trawl boards and anchors should be designed to minimise the probability and consequences of accidental damage.

The following should be considered in evaluating the form and level of lateral impact protection:

- Behaviour of fishing gear, anchors and dropped objects
- The need to provide structural support to protective roof structures
- Deflection of trawl boards and avoidance of anchor snagging by inclined or parabolic sides
- Avoidance of lateral impacts by barging or part-barging the production system (only normally considered where iceberg protection is necessary)
- The need to provide access to flowlines, umbilicals and other equipment

3.3.5 Dropped-Object Protection

A dropped object is any object with the potential to cause death, injury or equipment/environmental damage that falls from its previous static position under its own weight. Practical protection against dropped objects can be provided up to an impact energy of approximately 50 kJ (SJR 1994). Protection against dropped objects can take a number of forms, such as

- Increasing the protection
  - Installing local protection structures over Christmas trees and other sensitive areas
  - Installing total protection structures over the entire production system
- Introducing safety distances
  - Moving the rig a few metres off location before handling heavy objects (e.g. blowout preventers, or BOPs) and running them just clear of the seabed
  - Withdrawing the rig from the area when lowering the BOP
- Introducing safe areas
  - Not allowing activity of a certain kind within a specified area (e.g. trawling nearby platforms)
  - Locating the wellheads outside the range of dropped objects from a rig (e.g. satellite wells)
- Combining all of the above listed points

Any special dropped-object protection should be evaluated on the basis of probability of drop/hit combined with the hazardous consequence of hit (NORSOK U-001).

In areas with fishing activity, two options exist:

a. Establishment of a restricted zone (no bottom gear fishing is allowed in the area). This will require trawl resistant structures and/or continuous surveillance.
b. If the establishment of a restricted zone is not allowed, the use of overtrawlable structures is required.

Also, secondary securing devices (SSDs) secure a component at height in case the primary securing method fails. This could be a secondary safety wire (SSW), a safety net, and so on.

To evaluate the form and level of impact protection against dropped objects, the following points should be considered:

- Damage behaviour from dropped objects
- Interruption of maintenance and workover operations by protective covers
- Weight, cost and complexity of massive, retrievable protective covers
- Risk to the production system in running and retrieving protective covers
- Rig costs for moving off location during running and retrieving of heavy objects
- Rig and diver/submersible costs for removing small dropped objects
- Flow and control line costs associated with tying in satellite wells located outside the range of dropped objects to the adjacent rig

3.4 FABRICATION AND LOADOUT CONSIDERATIONS

Loadout is the operation of bringing the object (module, jacket, deck) from the quay onto the transportation barge.

Depending on the size and nature of the subsea structure, and any constraints imposed by the layout of the fabrication site, loadout may be achieved in the following ways (SJR 1994):

- Lifting using slings or spreader bars
- Jacking onto trailers for transport to the cargo barge
- Skidding on rails using greased sliding blocks and winches (although this is not common for the usual size of subsea structures)

3.5 INSTALLATION

The installation of the offshore platform will usually be performed by a vessel with an adequate crane on board. A heavy lift vessel (HLV) or diving support vessel (DSV) may be used. The following general rules apply:

a. Always use three or more slings when lifting structures.
b. Check for added mass and dynamic effects. Practically, a dynamic amplification factor of 1.5 is often used for analysis and design (SJR 1994; Lloyds 2014).
c. Skew loads during lift due to out of tolerances of lifting ropes are required by codes to be applied in the ratio of 75:25 (GPB 2010;

Lloyds 2014). Flexibility analysis of the structure may be presented to the certifying authority.

d. Lloyd's and DNV rules for transportation and installation are recommended.

e. Use 'diver- and ROV-friendly' designs to reduce vessel time offshore.

f. Compare the cost and schedules for grouted and swaged pile connections.

g. Assess the requirement for a rigging platform on top of the structure (a roof on a structure could possibly double as a rigging platform).

The following general methods of transportation and installation are available (SJR 1994):

a. Transportation by cargo barge, direct lift and lowering by crane barge or DSV

b. Transportation by multiservice vessel with on-board crane

c. Transportation by cargo barge and lowering by draw-works of drilling vessel or jack-up

d. Tow of floating structure to installation site, de-blasting and lowering by drilling vessel, jack-up or crane barge

3.5.1 Launch Truss

A launch truss is a very important vital component used for installation of jacket structures. Sometimes jacket structures are very large and cannot be lifted even with large cranes. Permanent structures such as launch trusses are provided on one side of the jacket to facilitate the loading out to the barge. If the jacket is designed for buoyancy, the jacket is launched into the sea after reaching its destined position for natural appending and levelling. When the jacket is launched, it floats because of its buoyancy. The jacket legs are sequentially flooded to make it upright, which is called appending (Lloyds 2014).

3.6 INSPECTION, REPAIR AND MAINTENANCE

3.6.1 Inspection

There are three types of underwater inspections:

a. *General Visual Inspection*

   The aim of this type of inspection is to assess the conditions in order to assess the platform integrity. An ROV may be used for the general visual inspection. The ROV visually inspects the whole structure underwater and takes cathodic protection potential measurements and marine growth measurements. The ROV uses a TV camera, potential measurement equipment and marine growth measurement equipment for this inspection. Any findings from the inspection are documented on videotape and still photos.

b. *Close Visual Inspection*

The area to be inspected is first cleaned thoroughly by an ROV using suction pads and steel brushes or air/grit blasting to remove marine growth. The cleaned area is then visually inspected with a colour camera.

c. *Non-Destructive Testing (NDT)*

The ROV cleans the area to be inspected before magnetic particle inspection (MPI) is done. The diver performs the MPI from a DSV.

3.6.2 Repair and Maintenance

The maintenance philosophy selected for installation and intervention operations on an offshore facility has a significant effect on the structure's overall layout (SJR 1994).

The maintenance philosophy could be either diver assisted or diverless. A diverless maintenance philosophy may require the use of an ROV or a purpose-designed tool package to perform the task. For a diverless maintenance philosophy, the following alternatives to divers should be considered:

- ROV
- Drill string tools
- Track mounted tools
- Dedicated remotely operated tools
- Modular maintenance systems

Equipment which requires frequent maintenance/repair or which is liable to failure should be modularised to make replacement or repairing easy on the surface (SJR 1994).

It is much easier to reduce the size of structures/modules, if required, at a later date, than to increase the area required, especially at or near the fabrication stage. Therefore, when deciding on the layout, a pessimistic (but realistic) approach about the area required must be taken. Sufficient space, accessibility, visibility, location points and work platforms must be provided. The structure should be clearly marked so that divers and submersibles can easily identify areas and locations of the structure. The markings should be clear and designed to avoid being obscured by marine growth. Cosmetic paintwork should be applied to equipment or components to aid recognition by divers or ROV cameras.

Divers are limited by a water depth of 300 m; therefore, for a platform in water depths greater than 300 m, an alternative to diver-assisted installation techniques must be addressed (SJR 1994).

The structural design may need to take into consideration the subsea intervention requirements for

- Well workover
- Servicing of wellhead equipment
- Opening or closing manual valves
- Control system monitoring during testing
- Installation of marine risers and tiebacks (including insertion of tensioning equipment)
- Installation of retrievable manifold and control system components
- Insertion of equipment for valve retrieval/replacement
- Corrosion monitoring
- Orientation and level monitoring
- Production system monitoring during testing (e.g. at connectors)

Follow the requirements of API RP 2D for Platform Inspection, Repair & Maintenance.

3.7 MATERIAL SELECTION

Material selection involves picking an engineering material, either metal alloy or a nonmetal that is inherently resistant to fire and a particular corrosive environment and also meets other criteria.

3.7.1 Structural Steel

Offshore structures are generally constructed of structural steel. The standards used for these steel materials depend on the particular project. Example of codes or standards normally used are shown in Table 3.1.

The four grades of steel which can be used for fabricating plates, rolled shapes, tubular members, built-up girders and beams are mild steel, high-strength steel, special mild steel and special high-strength steel.

TABLE 3.1   Design codes for structural steel

STRUCTURAL COMPONENT	STRUCTURAL CODE
Tubes	BS 7191, EEMUA 150, API 5L[a]
Plates	BS 7191, EEMUA 150, BS 4360, ASTM
Sections	BS 4848, BS 4
Bolts	ASTM A320, Grade L7
	ASTM A194, Grade 7 nuts can be used for L7 bolts
Nuts	ASTM A193, Grade 4/S3
Spherical washers	Steel

Source: SJR. 1994. Subsea Structure Design. London: JP Kenny Group.
[a] Linepipe code. Dimensions of tubulars can be extracted from here.

L7 bolts are recommended subsea with cathodic protection (CP). In the splash zone and where CP cannot be guaranteed, alloy 625 fasteners should be used.

3.7.2 Concrete

Concrete is a hard mass formed from the combination of cement, water, aggregates and sometimes additives (Table 3.2). The cement bonds the materials together through hydration. Aggregates have an important effect on concrete strength by providing rigidity to the material, which governs resistance to applied loads and undesired deformations.

Do not place concrete in deep water; concrete should be precast and then installed. However, if an admixture such as hydrocrete is added to the concrete mixture, it is possible to place the concrete directly in water. Hydrocrete is a water-resistant admixture used for surface treatment, protecting, waterproofing and repairing concrete and masonry.

It is important to use high-grade concrete with strength in excess of structural requirements. It is advisable to use sulphate-resisting cement and chloride-free additives to prevent the adverse effects of seawater on reinforced concrete. Chlorides may promote corrosion of steel. Concrete should not be cured at below 0°C or above 30°C.

Approval of concrete constituents is based on material testing where chemical composition, mechanical properties and other specified requirements are checked against a standard and other approved specification (e.g. DnV-OS-C502).

Simple thermal fire-response analyses may also be carried out for concrete partitions in order to determine the unexposed side surface temperatures (fire wall). In such analyses, it is essential to use the correct thermal material properties.

Protection against spalling consists of using a passive fire protective material on the surface of the element. The protective layer will reduce the

TABLE 3.2   Constituent materials for structural concrete

MATERIAL	DESCRIPTION
Cement	Sulphate-resisting cement conforming to BS 4027, ASTM C150 The following types of cement may be considered: Portland cement Portland composite cement Blast-furnace cement with high clinker content Pozzolanic cement Composite cement
Aggregates	Aggregates must be clean, natural stone, of sufficient strength, chemically inert, strong, hard, durable, of limited porosity, free from impurities that may cause rusting of steel reinforcement and resistant to decomposition when wet Aggregates should be evaluated for risk of an alkali silica reaction (ASR) in concrete
Fine aggregates	May be natural or derived from crushed stone. Sand or fine aggregates should be less than 5 mm in size. Fine aggregates shall comply with the requirements of BS 882
Coarse aggregates	Crushed natural stone with a maximum aggregate size of 20 mm. Heavy aggregates such as iron ore may be required for the production of extra heavy concrete
Water	Water must be clean and free of impurities that may impair the strength of the concrete. Salt water or seawater from excavations should not be used for mixing or as curing water for structural concrete. The fresh water used for mixing and curing the concrete specimens should be drinking water according to ASTM C190
Additives	Additives or admixtures to concrete mix should be used with caution and in accordance with the manufacturer's recommendations, laboratory test report, and international test standards Admixtures normally used are Plasticisers for increasing workability (e.g. hydrocrete for underwater concreting) Accelerators for reducing setting time Retarders for increasing setting time Calcium chloride admixtures should not be used for reinforced concrete. Admixtures should be delivered with a works certificate containing relevant chemical and physical properties

temperature gradients across the concrete surface and hence reduce the internal pore pressure in the concrete. For concrete exposed to seawater, the minimum grade is C35. For concrete which is not directly exposed to the marine environment, the concrete grade shall not be less than C25. Prestressed, reinforced concrete structures shall not be designed with a concrete grade less than C30.

3.7.2.1 Concrete Mix Design

The concrete composition should satisfy the requirement of DnV-OS-C502, BS8110 and the particular project specification. BS 8110 requires that concrete to be used in seawater has a compressive strength of 45 N/mm^2 (unreinforced) and 50 N/mm^2 (reinforced) and has 28-day characteristic crushing strength. The recommended concrete mix designs for offshore structures are as follows:

- Characteristic strength at 28 days should be 50 N/mm^2.
- Minimum cement content should be 300 kg/m^3. In the splash zone, the minimum cement content should be 400 kg/m^3. For reinforced or prestressed concrete not within the splash zone, the cement content is dependent on the maximum size of aggregate, as follows:
  - Up to 20 mm aggregate requires a minimum cement content of 360 kg/m^3.
  - 20 to 40 mm aggregate requires a minimum cement content of 320 kg/m^3.
  - For 40 mm and greater aggregate, the minimum required cement content shall be established by appropriate testing.
- Maximum water-to-cement ratio should be 0.45. In the splash zone, the ratio shall not be higher than 0.40.
- Maximum total chloride content of mix should not exceed 0.10% of the weight of cement in ordinary reinforced concrete and in concrete containing prestressing steel or as a relationship between chloride ions and weight of cement in mix.
- The coefficient of permeability of concrete should be low (10^{-12}–10^{-8} m/sec).

Concrete strength should be monitored by crushing test cubes of concrete samples taken during concreting. The strength of cured concrete should be assessed by taking core samples and crushing them. A rebound (Schmit) hammer may be used. Anti-crack reinforcement control can help solve cracking caused by heat of hydration of concrete, creep, shrinkage and thermal expansion.

To increase the resistance against attacks from salts in the seawater, cement with a moderate C_3A content may be used.

3.7.2.2 Reinforced Concrete

The components of a reinforced concrete are cement, fine sand, aggregates, water, admixtures and reinforcement, mixed to get a uniform matrix. Bar reinforcement is used to provide tensile strength to concrete. It may be mild or high-tensile steel; the modulus of elasticity of these steels is similar.

When concrete reinforcements strain because of tension, cracks in the concrete can develop and, consequently, water reaches the reinforcement, which causes corrosion.

The control of cracking is essential. Cracking can be controlled either by limiting the reinforcement stresses by having close reinforcement spacing or by prestressing. Prestressing steel shall comply with ISO 6934 and other relevant international standards. Prestressing steel shall be delivered with a works certificate.

The use of mild-steel reinforcement may be preferable to control cracking. If high-tensile reinforcement is to be used, the bars should be of high-bend deformed type.

The minimum concrete cover for reinforcement should be 50 mm. Greater covers may require additional considerations.

Reinforcement should comply with the BS4360 specification for weldable structural steels or similar. It should be thoroughly clean before pouring of the concrete commences.

3.7.2.3 Grout

Grout is a particularly fluid form of concrete used to fill gaps. Grout is made by the addition of water to cement or cement and fine aggregate. The constituents of grout are cement, water and often admixtures; fine aggregates may also be included. The mix is generally cement plus water.

Cement grout is used as follows:

- To act as ballast—Grout density may be taken as 20 kN/m³. Grout may be inserted into structural tubing of a frame or grout bags anchored to a structure.
- Shallow foundations—These may be required for levelling purposes or for providing a footing on the seabed. They consist of grouted bags or mattresses; the density and strength of the grout is relatively unimportant.
- Piling—Grout has two uses when piling:

- To bind a pile to the soil in a predrilled hole.
- To connect a pile to the leg of a structure. Grout is extensively used to 'cement' the annulus between the pile leg and jacket sleeve. An annular gap of 50–100 mm is usually selected. Fly ash may be used to replace part of the cement in order to reduce heat of hydration.
- Strengthening tubes—A tube subjected to a heavy impact load may be crushed locally. However, grout placed in the tube at the load point or in a dented tube will provide stability to the tube walls and will distribute a point load.

Grout strength depends on the design requirement and the grout design, but it is suggested that structural grout shall have a characteristic compressive strength higher than 35 MPa. Structural grout may be pre-packed blended or neat cement grout (SJR 1994).

Grout should be placed by pumping or injecting. This precludes the use of aggregates and is therefore termed cement grout.

Placement by pumping requires the preset grout to be of high fluidity and have a setting time suited to the time taken for placement. To achieve this, additives such as plasticisers and retarders, higher water-to-cement ratios or special cement may be used.

The mixing of grout should take between 1 and 4 minutes depending on the type of cement and the additives used. It should be placed within 30 minutes unless it incorporates a retarder.

After having completely filed all voids, grouting should be continued until the density of the grout issuing from the free end of the work is the same as that injected.

Ordinary Portland cement is frequently used, but the type of cement used depends on the required characteristics.

The water-to-cement (WC) ratio should not exceed 0.44. For neat cement, the WC should be about 0.40, and with admixes, the WC may be about 0.35.

3.8 CORROSION PROTECTION

Corrosion protection should be designed in accordance with NACE RP-0176. There are four methods of corrosion protection:

a. Material selection
b. Coatings (fire-resistive coating or intumescent coating)
c. Cathodic protection
d. Chemical inhibitor

For most applications, coatings and cathodic protection are used together. This ensures against coating breakdown late in the life of the system and failure of elements of cathodic protection system at any stage. It reduces the demand on the cathodic protection system, which enables economies to be made in its design. But where the design life is short, or where some corrosion can be tolerated, a single protection system may be appropriate.

Protection against corrosion may be achieved by the provision of painting and/or sacrificial anodes. Both protections have a life span that depends on the quality and quantity. Good-quality painting can offer protection for up to 10 years, but it is susceptible to local damage. Thick-film epoxy intumescent coating can provide the needed corrosion protection.

Rubber can also be used provide the corrosion protection in addition to environmental and impact protection, all while maintaining the required fire protection rating. For example, elastopipe, a rubber-based fire deluge system, can be used on an offshore platform.

Practically, most subsea pipelines are cathodically protected using sacrificial anode bracelets of the half-shell or segmented type attached to the pipeline at regular intervals. For subsea structures and platform legs, cathodic protection is by the use of slender standoff sacrificial anodes or flush-mounted sacrificial anodes which are welded to the structure. The three main metals used as sacrificial anodes are magnesium alloy, zinc alloy and aluminium alloy.

A corrosion allowance may be added to the specified wall thickness and the wall subsequently painted with a fire-resistive coating.

Structural steels corrode in seawater at a rate of approximately 0.1 mm per year, whereas the corrosion in mud may be as low as 0.01 mm per year. Therefore, a typical increase in thickness of a flat member with a required life of 10 years in water would be approximately $10 \times 0.1 \times 2$ *sides* = 2 mm.

A design life of 20 years is normally used for the cathodic protection design. The steelwork in the splash zone is usually protected by a sacrificial wall thickness of 12 mm to the members.

3.8.1 Service Life and Corrosion Protection

The structure must be designed for the life of the production system, usually for 20 years. Corrosion effect is likely to be the most significant factor affecting service life; however, if the structure is subject to cyclic loading, the fatigue life of structural steelwork elements should be considered (SJR 1994).

Structural components can be categorised when considering designs to ensure satisfactory service life as follows:

Primary areas:	Essential components of the production system (e.g. valve stools, manifold piping and intermediate supports, flowline connections and spool pieces, conductor guides)
Secondary areas:	Elements of template structure which, if subject to excessive corrosion or fatigue effects, could cause sag or collapse of primary areas (e.g. structural subframe, flow and control line porches, conductor guide supports, impact protection)
Other areas:	Other structural and nonstructural members

Considerations affecting corrosion protection design are

- Whether or not tubular members are flooded or sealed (this may depend on installation considerations)
- Whether or not service design life includes a final period during which corrosion is assumed to be acceptable
- Location of anodes to avoid interference with installation and maintenance/workover operation
- Location of anodes to avoid inhibition of efficiency due to buildup of drill cutting
- Temperature effects of production on corrosion rates and paint films
- Buildup of marine growth on paint-films, especially in areas artificially warmed by production

3.8.2 Cathodic Protection System

Corrosion protection of subsea structures is normally achieved by the use of anodes and often paints. In addition, on certain equipment items, a corrosion allowance may be added to the specified wall thickness and the wall subsequently painted (SJR 1994).

In each case, the paint acts to significantly reduce corrosion rates over a number of years. The choice of cathodic protection system is therefore a trade-off between

- Painting costs
- Anode costs or corrosion-allowance costs
- Weight control

Painting may, however, be necessary or desirable in certain areas for other reasons. The cost of this painting may thus be subtracted from the overall corrosion-protection cost.

The other reasons for painting may include

- Reducing corrosion effects during fabrication or storage
- Ensuring service life of essential components
- Protecting areas where pitting or crevice corrosion is likely to be a problem
- Discouraging marine fouling
- Aiding diver/ROV visibility

Structural Analysis

4

This chapter details the structural analysis of the offshore platform. Currently, there are software programs, such as SACS, FASTRUDL, MARCS, OSCAR, StruCad and SESAM, used to model and analyse an offshore platform. Complete structural analysis and design follows the selection of member sizes.

In the offshore industry, the design of fixed-steel offshore structures, typically known as jacket platforms, has been based on the API RP2A-WSD standard. The Working Stress Design (WSD) method is based on a safety factor provided only to the resistance of the material.

In this book, the limits state equations are based on the formulations provided in ISO19902 codes of design.

4.1 SELECTION OF MEMBER SIZES

Most expert engineers will be able to predict the preliminary platform structure and member sizes based on past projects and their experiences. The selected member sizes should then be verified and, if need be, resized through detailed analysis.

First, select the pile outside diameter. The deck leg outside diameter will generally be equal to the pile outside diameter. Once the main, pile, jackets and deck leg sizes are known, major jacket bracing, deck trusses beams and plating sizes can be selected.

Complete structural analysis and design follows the selection of member sizes.

4.1.1 Selection of Pile Size

Determine the maximum axial and shear forces on the pile. Select the outside diameter of the pile (commonly used pile outside diameter is 36–72 inches). The minimum pile outside diameter selected should be based on geotechnical properties such as soil strength, pile driving equipment capability and loads imposed on the pile. Calculate the pile penetration depth using the maximum pile axial load and a safety factor as in Equation 4.1.

$$Q_d = Q_f + Q_p = \int_{Z=0}^{L} f(Z)A_s(Z)dZ + q \cdot A_p \qquad (4.1)$$

where

Q_d = pile penetration depth
Q_f = total pile shaft skin friction resistance
Q_p = total pile end-bearing resistance
$A_s(Z)$ = side surface area of the pile per unit length at depth Z
q = unit end-bearing capacity (force/unit area)
A_p = gross end area of the pile
L = pile length
$f(Z)$ = unit skin friction capacity (force/unit area) at depth Z

Use a pile safety factor of 2.0 for the operational loading and 1.5 for the extreme design environmental loading case (refer to API RP 2A).

Select the pile wall thickness based on the axial load and bending moment acting on the critical pile cross-section. The pile wall thickness can be selected from tables in API RP 2A.

The minimum piling wall thickness used should not be less than

$$\left. \begin{array}{l} t = 0.25 + \dfrac{D}{100} \\[2mm] \text{Metric formula} \\[2mm] t = 0.25 + \dfrac{D}{100} \end{array} \right\} \qquad (4.2)$$

where

t = wall thickness (mm)
D = diameter (mm)

To increase resistance of pile to horizontal force, the following should be adhered to:

a. Increase pile diameter
b. Increase pile wall thickness
c. Increase the number of piles in the platform

d. Increase the lateral restraint of the piles within the jacket legs by fastening or locking the pile to the jacket leg at numerous points above the mud line or by filling the annulus between the pile and jacket leg with a material which will restrict or prevent movement of the pile within the jacket leg (fill to a short distance above the mud line)

4.1.2 Selection of Deck Leg Size and Deck Structure Dimensions

Determine the deck elevation relative to the sea surface before selecting the deck member size. The deck elevation must provide adequate gap (air gap) between the wave crest and the deck structure. Refer to API RP2A for the recommended air gap and any additional allowance for any predicted long-term seafloor subsidence.

The layout of the deck depends on the type of hydrocarbon processing to be performed on the platform.

The area required for the equipment, piping and cable routings, the vertical clearance and the access/egress requirements determine the deck area (see Figure 4.1) and deck elevations (ESDEP 2017).

The elevation on the lowest decks depends on the environmental conditions. The elevation of the cellar deck, that is the lowest deck, is based on the maximum elevation of the design wave crest, including tide and storm sway, plus a minimum air gap of 1.5 m, but should be increased if reservoir depletion will create significant subsidence. The vertical distance between the decks of the topside is generally in the range of 6–9 m in the North Sea.

Consideration of the prevailing wind direction is very important in determining the position of various components on the platform, such as the vent of the flare, cranes, helideck and the logistic and safety provisions (ESDEP 2017). The requirements for the various topside components should be based on API-RP 2G.

4.1.2.1 Deck Leg Size

For a jacket platform, the deck leg outside diameter will generally be equal to the pile outside diameter. For a tubular leg, the radius of gyration (r) of its cross-section can be calculated from 0.35D. This applies to thin wall cylinders. For nontubular cross-sections, this parameter can be picked from a table.

Initially select a buckling length factor (k) of 1.5 and then calculate the slenderness ratio (kL/r) of the deck leg, with L as the actual deck leg length. Calculate the allowable axial stress and maximum axial load and moment on the deck leg. Select the approximate wall thickness of the deck from a table. Calculate the axial and bending stress.

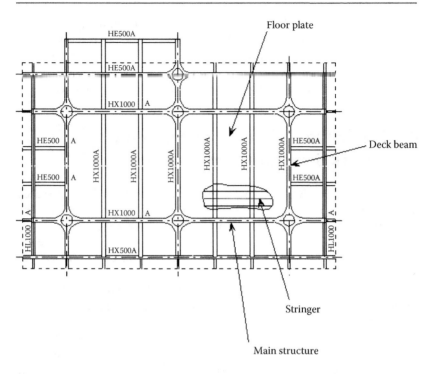

FIGURE 4.1   Basic structural grid for a jacket-based deck. (From ESDEP. n.d. *ESDEP WG 15A*. Retrieved July 10, 2017, from http://fgg-web.fgg.uni-lj.si/%7E/pmoze/ESDEP/master/wg15a/l0800.htm)

4.1.2.2 Deck Framing

   a. Select spacing between stiffeners (typically 500–800 mm).
   b. Derive the plate thickness from formulas, accounting for local plastification under the wheel footprint of the design forklift truck.
   c. Determine by straight beam formulas the sizes of the main girders under 'blanket' live loads and/or the respective weight of the heaviest equipment.

4.1.2.3 Selection of Deck—Main Member Sizes

Select the deck framing based on the deck span, operational requirements, the local construction practices and access clearance needed to install and maintain drilling and production equipment.

The deck beam sizes can be selected by using simple or continuous beam formulas.

When using a plate floor system, select the deck plate size by using an upper bound distributed deck load and selecting a deck plate thickness using the plate bending formulas.

The selection of the topside main structure concept, truss or portal frame, is linked with the decision of the position of the longitudinal structure in the cross-section. In a 20–25 m wide deck, trusses will generally be arranged in longitudinal rows: centre line and both outer walls.

In the Dutch sector of the North Sea, transverse column spacings are typically 9 m for a wellhead platform to 15 m for a production platform. Longitudinal spacings are typically 15 m (ESDEP 2017).

Typical span of structural items are

a. Floor plate: 1 m
b. Stringer (longitudinal): 5 m
c. Deck beam (transversal): 15 m
d. Main structure (longitudinal): 15 m
e. Column

Deck beams supporting the floor panels or providing direct support to major equipment are generally provided as HE 800–1000 beams, though HL 1000 (400 mm wide) or HX 1000 (450 mm wide) are also used for heavier loads or greater spans (ESDEP 2017). For offshore applications, fire classified decks such as H120 and H60 steel decks and bulkheads can also be used.

The major joint in the deck beam is that with the main structure. The joint configuration is strongly determined by the prefabrication concept and elevation of the flanges. It is different for the stacked and for the flush concept (ESDEP 2017).

4.1.3 Selection of Bracing Size

The floor plate is preferred to act as horizontal bracing. If, however, separate bracing members are required, the elevation must be chosen carefully. The bracing members have to pass with sufficient clearance under the stringers and penetrate the web of the deck beams at sufficient distance from the lower flange. They also require good access for welding of the joint. These requirements generate the elevation and the maximum feasible diameter of the brace.

Horizontal bracing can easily clash with vertical piping and major hatches. Assembly of the braces is generally quite cumbersome.

Choose a brace member diameter with a slenderness ratio (kL/r) in the range of 70–90. This is an industry-accepted practice. Provide additional support points such as the use of lateral braces or X-bracing to lower the buckling length

Verify the capacity of the joint the member will be framing into. For example, you can use brace-to-chord outside diameter ratios higher than 0.30 to improve the joint capacity. If the bracing is to be located in a splash zone, consider an additional corrosion allowance. Try to keep the pipe wall thickness modulus (*D/t*) in the range of 19–30 (Figure 4.2).

4.1.3.1 Jacket Bracings

The following selection method is in accordance with the requirement of European Steel Design Education Program (ESDEP).

a. Select the diameter in order to obtain a span-to-diameter ratio between 30 and 40.
b. Calculate the axial force in the brace from the overall shear and the local bending caused by the wave assuming partial or total end restraint.
c. Derive the thickness such that the diameter-to-thickness ratio lies between 20 and 70 and eliminate any hydrostatic buckle tendency by imposing $d/t < 170/^3\sqrt{h}$ (*h* is the depth of member below the free surface).

4.1.4 Equipment and Living-Quarter Modules

Equipment modules (20–75 MN) have the form of rectangular boxes with one or two intermediate floors. The floors are steel plates (6, 8 or 10 mm thick) for the roof and lower floor and grating for intermediate floors (ESDEP, n.d.f).

In living-quarter modules (5–25 MN), all sleeping rooms need windows, and a number of doors must be provided in the outer walls. This requisite can have an extreme effect on the truss arrangements. Floors are flat or stiffened plates (ESDEP 2017).

There are three types of structural concepts, all avoiding interior columns, can be distinguished:

a. Conventional trusses in the walls
b. Stiffened plate walls (so-called stressed skin or deckhouse type)
c. Heavy base frame (with wind bracings in the walls)

(a)

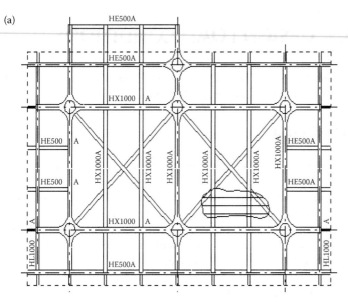

(b)

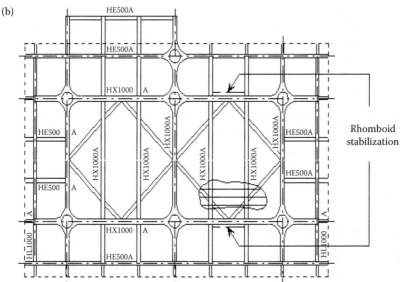

Rhomboid
stabilization

FIGURE 4.2 Alternative configurations for underfloor bracings. (a) Crossed, not very feasible for upper deck due to interference of brace/column joint with padeyes or trunnion stiffeners and (b) rhomboid, more feasible for upper deck, however requires stabilization. (From ESDEP. n.d. *ESDEP WG 15A*. Retrieved July 10, 2017, from http://fgg-web.fgg.uni-lj.si/%7E/pmoze/ESDEP/master/wg15a/l0800.htm)

4.2 DESIGN LOAD CONDITIONS

A load condition is a specific type of loading (e.g. dead load or live load). All the load conditions detailed below should be considered.

4.2.1 Dead Loads

Dead loads are the self-weight of the structural framework of the platform and its fixed, permanent and fabricated fittings (e.g. grout, piles, ballast, supports, anodes, pipework, etc.), weight of equipment and associated structures permanently installed on the platform and hydrostatic loads due to water depth on sealed hollow members of a submerged structure (API 2000; Kövecsi 2014). Hydrostatic forces include buoyancy hydrostatic pressure. Hydrostatic loads should not be applied to free-flooding members.

4.2.1.1 Equipment Loads

The weight of equipment may be obtained from a vendor. The weight of the equipment shall be distributed to the deck beams or plating depending on the load-transfer method adopted for the design of the equipment skid (Kharade and Kapadiya 2014).

4.2.1.2 Fluid Loads

The fluid loads are based on equipment operating weight and can be obtained from the equipment manufacturer. The fluid load is calculated as shown in Equation 4.3.

$$W_{fluid} = W_{oper} - W_{dry} \tag{4.3}$$

where W_{oper} and W_{dry} are the weight of equipment in operating and dry conditions, respectively.

4.2.1.3 Drilling Loads

Drilling loads include the weight of drilling equipment such as rig, drill strings and mud tanks. The drilling rig is applied as a point load because it is not fixed equipment (Kharade and Kapadiya 2014).

4.2.2 Live Loads

Live loads will be experienced during the platforms service life due to installation, drilling, production and maintenance operations (API 2000). Examples are

a. The weight of drilling and production equipment which can be added or removed from the platform.

b. The weight of living quarters, heliport and other life-support equipment, life-saving equipment, diving equipment and utilities equipment which can be added or removed from the platform.

c. The weight of consumable supplies and liquids in storage tanks.

d. The forces exerted on the structure from operations such as drilling, material handling, vessel mooring and helicopter loadings.

e. The forces exerted on the structure from deck crane usage. These forces are derived from consideration of the suspended load and its movement as well as dead load.

f. Flow and control line pull-in and connections.

g. Landing and retrieval of Christmas trees, BOPs and other retrievable items.

h. Thermal expansion effects of flowline connectors, pipework, and so forth, due to hot-producing wells.

i. Pilling activities during installation phase.

j. Riser loads.

k. Casing installation and setting.

4.2.3 Environmental Loads

Environmental loads are imposed on the platform from the surrounding environment. The environmental load to be considered includes hydrodynamic loads, wind loads, current, snow, ice loads, earth movement and earthquake loads (Kharade and Kapadiya 2014; API 2000).

The API specifies minimum design criteria for a 100-year design storm.

4.2.3.1 Buoyancy Loads

Buoyancy loads are loads experienced by a submerged structure due to the action of the displaced water acting through the centre of buoyancy of each member. The members that are to be flooded will have to be identified and considered.

4.2.3.2 Wind Loads

Wind loads are assessed according to API RP 2A-WSD and BS 6399: Part 2 (BS CP3: CHV: Part 2). Wind loads act on the portion of a platform above the water level, as well as on any equipment, housing, derrick, and so on, that are located on the deck. The time interval over which wind speeds are averaged is a very important parameter.

4.2.3.2.1 Wind Velocity
The wind velocity profile can be obtained from API RP 2A and the wind
velocity is calculated as

$$\frac{V_h}{V_H} = \left(\frac{h}{H}\right)^{1/n} \tag{4.4}$$

where
  V_h is the wind velocity at height h
  V_H is the wind velocity at reference height H, typically 10 m above mean
    water level
  $1/n$ is 1/13 to 1/7, depending on the sea state, the distance from land and
    the averaging time interval. It is approximately equal to 1/13 for
    gusts and 1/8 for sustained winds in the open ocean

4.2.3.2.2 Static Wind Force
The static wind force acts perpendicular to an exposed area and is computed
as follows

$$F_w = \frac{1}{2}\rho V^2 C_s A \tag{4.5}$$

where
  F_w = static wind force (N)
  ρ = the wind density ($\rho = 1.225$ kg/m³)
  A = exposed area (m²)
  C_s is the shape coefficient ($C_s = 1.5$ for beams and sides of buildings,
    $C_s = 0.5$ for cylindrical sections and $C_s = 1.0$ for total projected
    area of platform)

4.2.3.3 Hydrodynamic Loads

Hydrodynamic loads are generated by the movement of water particles
around a submerged object, vessel motion, wave-induced and steady-state
currents, and so forth. Hydrodynamic loads such as inertial force, drag
force, lift force and flow-induced cyclic loads need to be calculated using the
Morison's equation (API 2000). Hydrodynamic force is the summation of the
inertial force and drag force.

$$F = \rho C_m V u' + \frac{1}{2}\rho C_d A u |u| \tag{4.6}$$

$$F = \rho C_m V u' + \frac{1}{2} \rho C_d A u^2 \tag{4.7}$$

$$F = C_m \frac{\rho \pi D^2}{4} u' + C_d \frac{\rho D}{2} u^2 \tag{4.8}$$

As a rule, Morison's equation may be applied when $D/L \le 0.2$, where D is the member diameter and L is the wavelength. Any of Equations 4.6 to 4.8 can be used to calculate the hydrodynamic load.

4.2.3.3.1 Inertial Force
Inertial force is caused by acceleration of water particles under wave action. It acts horizontally against the platform. Inertial force is calculated as

$$F_I = \rho C_m V u' \tag{4.9}$$

$$F_I = C_m \frac{\rho \pi D^2}{4} u' \tag{4.10}$$

where
$\quad F$ = total wave force on the object
$\quad F_I$ = the inertial force
$\quad \rho$ = the density of the surrounding water
$\quad D$ = the overall outside diameter of member
$\quad u'$ = the wave-induced particle acceleration (flow acceleration)
$\quad C_a$ = the added mass coefficient
$\quad C_m = 1 + C_A$
$\quad C_m$ is the inertia coefficient determined from the appropriate design code
$\quad V$ is the volume of the body

4.2.3.4 Drag Force

Drag force is caused by the velocity of water particles under wave and current action. The drag force is calculated as shown in Equations 4.11 and 4.12.

$$F_D = \frac{1}{2} \rho C_d A u |u| \tag{4.11}$$

$$F_D = \frac{1}{2} \rho C_d A u^2 \tag{4.12}$$

where

F_D = drag force
ρ = mass density of surrounding water
u = flow velocity relative to the object
A = reference area or cross-sectional area
C_D = drag coefficient (it is dimensionless)

The values of C_d and C_m depend on the wave theory used, surface roughness and the flow parameters. $C_d = 0.6$–1.2 and $C_m = 1.3$–2.0 (refer to API RP 2A). Additional information can be found in the DNV rules.

4.2.3.5 Lift Force

The lift force acts vertically upwards and is caused by the velocity of water particles passing over the top of the member under wave and current action. Lift forces can be estimated by taking $C_L = 1.3\, C_d$.

The lift force is calculated as

$$F_L = \frac{1}{2}\rho C_L D u^2 \tag{4.13}$$

where

F_L = lift force (N)
D = overall outside diameter of member (m)
C_L = lift coefficient as determined from the appropriate design code
u = velocity due to design waves (m/s)
ρ = water density (kg/m^3)

All other parameters are as defined above.

4.2.3.6 Slamming Force

Slamming forces acting on the underside of horizontal members near the mean water level are impulsive and nearly vertical (ESDEP 2017).

$$F_s = \frac{1}{2}\rho C_s D u^2 \tag{4.14}$$

where

F_s = slamming forces (N)
C_s = slamming coefficient as determined from the appropriate design code
ρ = water density (kg/m^3)

All other parameters are as defined above.

4.2.3.7 Ice and Snow Loads

Ice shall be a major problem for offshore platforms in the arctic and subarctic zones, as it can generate horizontal and vertical forces through its formation and expansion (ESDEP 2017). The Eurocode provides detailed information on snow loads.

Horizontal ice forces can be estimated as follows:

$$F_i = C_i f_c A \qquad (4.15)$$

where
F_i = the horizontal ice force
A = the exposed area of structure
F_c = the compressive strength of ice
C_i = the coefficient accounting for shape, rate of load application and other factors, with usual values between 0.3 and 0.7

4.2.4 Accidental Loads

Accidental loads or impact loads may be experienced during the platform service life. An example is an impact caused by a barge or boat berthing against the platform or by drilling operations, collision with the platform, dropped objects, anchors, fire, blast, unintended flooding, extreme weather, trawl board impact in any lateral direction (typically 20 kN at 3 m/s), trawl snagging (50 tonnes point load or 100 tonnes overall), designed accidental load due to a dropped object acting in a vertical or inclined direction, designed accidental pull (50 tonnes) in any direction due to snagged anchor, loss of internal pressure necessary for maintaining minimum required safety level, and so on. Such accidental loads are likely to be critical to the structural integrity of the structure (SJR 1994; API 2000).

Accidental loads should be considered in conjunction with environmental loads (Figure 4.3).

4.2.4.1 Fire as a Load Condition

Fire is a combustible vapour or gas that combines with an oxidiser in a combustion process that is manifested by the evolution of light, heat and flame (Paik 2011).

The British Standard BS 5950 Part 8 (BSI 1990b) and Eurocode 3 Part 1.2 (CEN 2000b, hereafter referred to as EC3) are the main sources of information for fire-resistant design of steel structures. For structural fire-resistant design,

FIGURE 4.3 A platform damaged by storm. (Image reused with permission from Raabe, D. n.d. *Dramatic failure of materials in drilling platforms*. Retrieved August 12, 2017, from http://www.dierk-raabe.com/dramatic-material-failure/oil-drilling-platforms/)

reduced partial safety factors for structural loads should be used. Reduced load factors for the fire limit state design can be obtained from BS 5950 Part 8 (BSI 1990b) and Eurocode 3 Part 1.2 (CEN 2000a).

Determine relevant fire sizes and loads to be applied for structural integrity analyses. This can be used as input to the structural calculation programs (e.g. FAHTS/USFOS) for advanced structural fire design (Lloyds 2014).

The British Standard adopts load ratio while, in Eurocode, load level is used. Hence, the load level at exposure time in fire condition can be expressed as given in Eurocode 3 and by Lawson and Newman as

$$\eta_{fi,t} = \frac{E_{f,d,t}}{R_d} = k_{y,\theta} \left[\frac{\gamma_M}{\gamma_{M,fi}} \right] \le \frac{R_{f,d,t}}{R_d} \tag{4.16}$$

where

$\eta_{fi,t}$ = load level at exposure time in fire condition
$E_{f,d,t}$ = actions effect at the fire limit state (N, Nm)
R_d = design value of resistance (N, Nm)
$R_{f,d,t}$ = reduced member resistance (design resistance in fire condition) (N, Nm)
γ_M = partial safety factor on materials

$\gamma_{M,fl}$ = partial safety factor on materials in fire condition
$K_{y,\theta}$ = reduction factor for yield strength with temperature, θ

Equation 4.16 is considered effective at the fire limit state.

4.2.4.1.1 Fire Limit State

In the limit state definition, the safety of structures is assured when

$$\sum \gamma_f E \leq \frac{R}{\gamma_M} \tag{4.17}$$

where
γ_f = the partial safety factors on actions
γ_M = partial safety factors on materials
R = member resistance (N, Nm)
E = Young's modulus (N/m²); actions effect (N, Nm)

At the fire limit state, the partial safety factors for permanent and variable actions are put to unity to account for the likelihood of reduced loading in the event of a fire, thereby leading to the design value of the actions adjusted as follows

$$E_{fi,d,t} = G_d + \varphi_1 Q_{k,1} + \varphi_2 Q_{k,2} \tag{4.18}$$

where
$E_{fi,d,t}$ = actions effect at the fire limit state (N, Nm)
φ_1, φ_2 = load combination factors depending on the limit state under consideration
G_d = permanent actions (N)
$Q_{k,i}$ = variable actions (N, $i = 1, 2,$)

4.2.4.2 Explosion Loads

Explosion is a sudden and violent release of energy, the violence of which depends on the rate at which the energy is released. The energy stored in a car tyre for example can cause an explosive burst, but it can be dissipated by gradual release.

Three basic types of energy released in an explosion are

a. Physical energy: Includes energy in gases, strain energy in metals, and so on
b. Chemical energy: Derived from a chemical reaction, usually combustion
c. Nuclear energy (not considered here): Chemical explosions and, particularly, gas phase explosions, which are of interest

Explosions may either be categorised as *deflagrations* or *detonations*. In deflagrations, the flammable mixture burns relatively slowly. Rates of flame propagations are characterised by burning velocity (Su), that is, the velocity of a plane flame normal to itself and relative to the unburned reactants.

A *deflagration* is an exothermic reaction (a moving flame front) which spreads from the burning gases to the unreacted material by conduction, convection and radiation. In this process, the combustion zone progresses through the material (flammable mixture) at a rate that is less than the speed of sound in the unreacted material (Harris 1983).

In contrast, a *detonation* is an exothermic reaction characterised by the presence of a shock wave in the material that establishes and maintains the reaction. A distinguishing characteristic of detonation is that the reaction zone spreads at a speed greater than the speed of sound (Farid and Longinow 2003).

Under proper conditions, flammable and combustible gases, mists or dusts suspended in air or another oxidant can burn when ignited (ASCE 1997; Farid and Longinow 2003). A detonation is an abnormal flame process in that no combustion device to date makes use of it, and few accidental occurrences seem to have involved it. Detonation occurs at many times the speed of sound in the unburned gas, to be specific, at the speed of sound in the hot products of combustion. Stable detonation is an accurately predictable phenomenon, and the theory is well established. This is dealt with elsewhere, but basically, detonation results from the passage of a strong shock wave through a combustible mixture causing its instantaneous compression, heating, ignition and reaction. The energy released by the reaction is fed to the shock wave and sustains it (Harris 1983).

For hydrocarbon-air mixtures, the detonation velocity is 2000–3000 m/s (note that the velocity of sound in air at 0°C = 330 m/s).

Detonations develop higher pressures and are much more destructive than deflagrations. The peak deflagration pressure in a closed vessel is about 8 bar; for detonations it may be over 20 bar.

Deflagrations may accelerate to detonations in long closed systems, for example, pipes. Acceleration to detonation appears to be unlikely or impossible in three-dimensional systems unless powerful initiators, for example, explosive detonators, have been used.

Blast or explosion loading may also be categorised as a *confined explosion* and an *unconfined explosion*. Unconfined explosions occur in the free air and may occur near the ground. Confined explosions occur in fully confined, partially confined and fully ventilated areas. Confined explosions occur inside the structure, and the combined effect of high temperature and accumulation of gaseous products of chemical reactions in the blast may lead to the collapse of the structure if not designed to withstand internal pressure.

A basic distinction may be drawn between explosions which occur in some sort of containment and those which occur in *unconfined* vapour or gas clouds.

In general, explosions which occur in buildings or plants are initially confined but, due to failure of part or whole of the structure, become at an early stage vented explosions. Our concern generally is with vented and confined explosions.

Often, accidental release of gas will occur in some form of containment provided by a building or section of industrial plant. This will lead to a *confined gas explosion*. Under conditions of complete confinement, most fuel gases can produce a maximum pressure rise of about 8 bar. Most buildings and heating plants are incapable of withstanding such pressures. However, internal gas explosions rarely cause complete destruction because, either by design or fortuitously, the pressure is relieved at an early stage of the explosion by failure of weak components (Harris 1983). Such explosions are termed *vented confined explosions* (Figure 4.4).

4.2.4.2.1 Hydrocarbon Explosion Load

A hydrocarbon explosion is a process in which combustion of a premixed hydrocarbon gas-air cloud causes a rapid increase of pressure waves that generate blast loading (Ali 2007). Common types of explosions include accidental

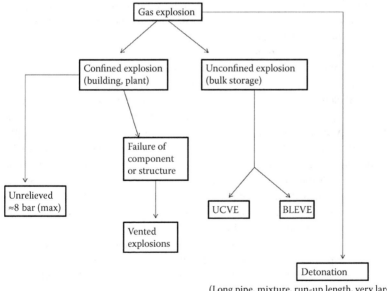

FIGURE 4.4 Confined, unconfined, vented explosions.

explosions resulting from natural gas leaks or other chemical/explosive materials. The three elements necessary for explosion to occur are

a. Oxidiser
b. Fuel
c. Ignition source

Gas explosions can occur inside process equipment or pipes, in buildings or offshore modules, in open process areas or in unconfined areas. The worst case that is possible to occur on the platform is a hydrocarbon explosion that generates blast loading. A topside structure which forms a percentage of the superstructure where most facilities of the process plant are sited must be protected against hydrocarbon explosion loading. Usually, safety aspects and prevention measures on offshore installations are very tough; hence, the likely incident of an explosion is reasonably low.

Hydrocarbons can explode through ignition when combined with an oxidiser (usually air). Thus, when the temperature rises to the point at which hydrocarbon molecules react spontaneously to an oxidiser, combustion takes place. This hydrocarbon explosion causes a blast and a rapid increase in overpressure (Paik 2011).

In an unconfined or free area, the pressure waves are released/discharged in all possible directions within the duration of a few milliseconds as a pressure impulse. The magnitude of overpressure for the unconfined area is lower than a confined/restricted area.

Most facilities are in-placed at the topside, which consists of structural members, piping, equipment, cables and other appurtenances that can hinder the free movement of these waves. Therefore, introducing congestion and confinement significantly increases the magnitude of overpressure loads (Ali 2007).

In the Piper Alpha incident (1988), where 167 men died, a major part of the installation was burnt down. A series of explosions destroyed the Piper Alpha oil platform in the North Sea. An inquiry blamed the operator, Occidental, for poor maintenance and safety procedures. The incident started with a rather small gas explosion in a compressor module that caused a fire which subsequently resulted in the rupture of a riser. The explosion ruptured fire walls that were designed to withstand oil fires, but not gas explosions, and were not retrofitted when the rig was modified to accept gas.

It is therefore a challenge for offshore structural engineers, especially for a new development project, that the effects of blast loadings should be taken into serious consideration from the onset of the design stage (Wikipedia 2017; Ali 2007).

4.2.4.2.1.1 Overpressure Load   The impact of overpressure from explosions and that of elevated temperature from fire are the primary concerns in

TABLE 4.1 Nominal blast overpressure

BLAST AREA	PRESSURE (BAR)
Wellhead deck	2.0–2.5
Gas separation facilities	1.5–2.0
Gas treatment compression facilities	1.0–1.5
Process area, large or congested	0.5
Process area, small and not congested	0.2
Open drill floor	0.2
Totally enclosed compartment	4.0

Source: Ali, R. M. 2007. *Performance-Based Design of Offshore Structures Subjected to Blast Loading.* London: Imperial College.

terms of the actions that result from hazards within the risk assessment and management framework (Paik 2011).

A *quantitative risk assessment* (QRA) study provides engineers with the nominal overpressure values, or alternatively the overpressure exceedance (probability of any particular level of overpressure being exceeded), for design purposes. A compilation of design overpressure is given in Table 4.1.

As shown in Figure 4.5, low overpressure values are dominated by a higher frequency of exceedance values and vice versa. The maximum peak pressure (at low frequency of exceedance) of 3.0 bar up to 4.0 bar is recommended for the design of primary supporting trusses, while nominal load for open deck flooring

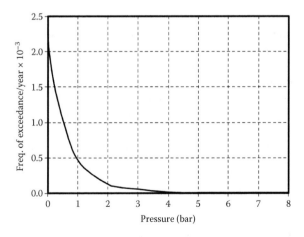

FIGURE 4.5 The frequency of exceedance–overpressure. (From Ali, R. M. 2007. *Performance-Based Design of Offshore Structures Subjected to Blast Loading.* London: Imperial College.)

is recommended between 0.5 bar and 1.0 bar (at high frequency of exceedance) (Ali 2007).

In a process module on an offshore platform, a large separator can block off venting across it, effectively behaving like a wall resulting in a large overpressure load. This effect can be significant even when confinement is minimal (Corr and Tam 1998).

4.2.4.2.1.2 Drag Load   A gas explosion on an offshore platform will generate drag loads in the offshore installations because of their high velocity affecting pipe work and vessels, possibly leading to damage to safety critical systems, loss of containment of hydrocarbon and escalation of the event.

Structural engineers have become aware of overpressure effects through working on the design of blast walls, steel work and consideration of escalation effects.

The Steel Construction Institute's 'Interim Guidance Notes' (SCI IGN) (Steel Construction Institute, 1992) are used extensively for reference and can be used to evaluate the structural performance of structural steel work under blast load (Corr and Tam 1998; IGN 1992).

The wind of the blast load causes drag loads (drag force), and the impact of these drag forces is significant on small objects such as electrical cable trays and piping. Drag force is calculated using Equation 4.19.

$$F_D = C_D \frac{1}{2} \rho A u_{gas}^2 \qquad (4.19)$$

where

F_D = drag force or drag load from the blast per unit length
ρ = density of gas
u_{gas} = flow velocity of gas
A = reference area or cross-sectional area (projected area of the object normal to the flow direction)
C_D = drag coefficient for blast (it is dimensionless)

Equation 4.19 can be expressed in terms of gas temperature as

$$F_D = \frac{1}{2} C_d \rho_{amb} \left(\frac{283}{T} \right) A u_{gas}^2 \qquad (4.20)$$

where

ρ_{amb} = the density of the air at 283 K
T = the temperature of the burnt gas (K)

All other parameters are as defined above.

The density of gas and velocity of gas flow are unpredictable, therefore Equation 4.19 is modified into a simplified expression as follows (Eknes and Moan 1994).

$$q_D = C_D p(t) D \qquad (4.21)$$

where

q_D = line load on pipe
$p(t)$ = maximum overpressure time-history
D = outer pipe diameter
C_D = drag coefficient (it is dimensionless)

In many types of topsides or decks, grating has been installed to improve ventilation and allow venting of possible explosion gases. Grating allows partial venting of the explosions, allowing some of the gases to pass through. The resultant loading on the grating is a combination of the overpressure differential across the grating and drag load created by the gases passing through the voids in the grating (Corr and Tam 1998).

Therefore, loading from a gas explosion on grating is mainly from drag. The drag loading on platform grating can be calculated assuming the following:

• Differential pressure across constituent elements of the grating (i.e. plates and cylinders) is negligible.
• The total load on the grating plate is the sum of drag load acting on individual constituent elements.
• Drag coefficients are similar to those measured under steady-state situations.
• Flow is turbulent.
• The aspect ratio of the constituent element length/D is large, where D is the diameter in the case of cylinders and the width in the case of plates.

The appropriate drag coefficient (C_D) for all the elements comprising the grating is 2.0. Drag loading on a grated deck is calculated as

$$Drag_{grating} = \frac{1}{2} C_d \rho A u_{gas}^2 \qquad (4.22)$$

where A is the cross-sectional area presented by the grating of the wind.

The drag loading for a square metre of grating is

$$Drag_{grating} = 0.4 \rho u_{gas}^2 \qquad (4.23)$$

4.2.4.2.1.3 Missile Load The consequent effect of the drag loads are missile loads. The rupturing of equipment causes all kinds of flying objects as well as fragments; these are considered missiles (Ali 2007). Based on Equation 4.19, the equation for calculating the missile load is generated as

$$F_{missile} = \frac{1}{2} C_d \rho A (u_{gas} - u_{missile})^2 \tag{4.24}$$

where
$F_{missile}$ = drag force on the missile
$u_{missile}$ = velocity of the missile

Tam and Simmonds (1990), proposed an equation for calculating the peak velocity using correlated maximum gas velocity ($u_{g\,max}$), duration of the gas flow ($t_{g\,max}$) and mass of the missile to empirical parameter 'α' as

$$\alpha^2 = \frac{1}{2} C_d \rho A u_{g\,max} \frac{t_{g\,max}}{M} \tag{4.25}$$

where
$u_{g\,max}$ = maximum gas velocity
M = mass of the missile
$t_{g\,max}$ = duration of the gas flow

The peak velocity of the missile is calculated as

$$u_{missile} = u_{g\,max} \left[1 - \frac{\sqrt{2}}{\alpha} \tan^{-1} \left(\tanh \frac{\alpha}{\sqrt{2}} \right) \right] \tag{4.26}$$

The Tam and Simmond's model was based on the following assumptions (Corr and Tam 1998):

a. The gas was burnt and
b. The velocity increased linearly to a peak and then decreased linearly with equal rise and fall times.

4.2.4.2.2 External Explosion Load from High Explosives

An explosion is a rapid release of stored energy characterised by a bright flash and an audible blast. Part of the energy is released as thermal radiation (flash), and part is coupled into the air as air blast and into the ground as ground shock, both as radially expanding shock waves.

The explosion load is one of the most important parameters to determine whether or not the structural integrity is satisfactory. Explosion load is an uncertain parameter because it depends on variable factors, such as maximum pressure, rise time and pulse duration.

Compared with gases, solids possess thermal conductivities which are typically 10 times as great and diffusivities which are many powers of 10 lower. This, coupled with their very high energy density, produces very energetic combustion, either detonation or deflagration, producing very high pressures (Harris 1983).

Detonating explosives are normally termed high explosives and may produce peak pressures of up to 300,000 atmospheres. Deflagrating explosives are termed low-explosives in the United States and produce peak pressures below 4000 atmospheres.

Of the high explosives, some are detonated by all normal ignition sources, for example, heat, spark, mechanical impact, and are termed primary explosives. Secondary explosives, on the other hand, will only detonate under influence of an externally applied shock or detonation wave and merely deflagrate when ignited by a flame. Small quantities of primary explosives are used in detonators to induce explosion in bulk secondary explosives.

High explosives produce high-amplitude, short-duration pressure profiles compared with gas explosions, and this is a weakness in the application of correlations based on high explosives to other explosive situations. Nevertheless, the TNT (trinitrotoluene) model is widely used for estimating the effects of explosions on plants and buildings (Harris 1983).

The procedure for calculating the blast or explosion load from high explosives such as TNT is as follows:

a. Calculate the weight of the charge, W, charge distance of the structure, R, charge height, H (for explosions in air) and structural dimensions.
b. Apply a safety factor of 0.20 (or 20%). Due to a variety of uncertainties, it is recommended to apply a safety factor to the charge weights and augment them by approximately 20%.
c. Divide the structure into sections or zones and calculate the explosion parameters for each section or zone.
d. Calculate the scaled charge distance, Z.
e. Calculate all other parameters as shown below.

Assume the blast occurs at a distance outside the platform and the entire structure is to be affected by the blast waves. The maximum overpressure is calculated to determine loads that can be exerted on the platform due to blasting. The proposed equations by Kinney and Graham, and the TM5-1300 diagram (see Figure 4.6) can be used to accurately estimate the overpressure due to blast (TM5 1990).

Using the Kinney and Graham suggestion for calculating maximum overpressure (P_s),

$$\frac{P_s}{P_0} = \frac{808\left[1+\left(\dfrac{z}{4.5}\right)\right]^2}{\left[1+\left(\dfrac{Z}{0.048}\right)^2\right]^{1/2} \cdot \left[1+\left(\dfrac{Z}{0.32}\right)^2\right]^{1/2}\left[1+\left(\dfrac{Z}{1.35}\right)^2\right]^{1/2}} \qquad (4.27)$$

where

P_0 = ambient air pressure (atmospheric pressure, bar)
P_s = peak static wave front overpressure (bar)
Z = scaled distance (m/kg$^{1/3}$), $z = R/(W^{1/3})$
R = the distance from detonation point to the point of registered pressure
W = the weight equivalent TNT charge (kg)

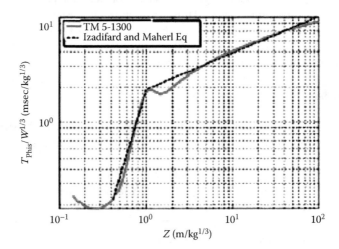

FIGURE 4.6 Comparisons between Izadifard and Maheri equation with TM5-1300 diagram. (From Abdollahzadeh, G. R. and Nemati, M. n.d. *Risk Assessment of Structures Subjected to Blast.* Retrieved August 28, 2017, from www.research-gate.net/publication/274509776_Risk_Assessment_of_Structures_Subjected_to_Blast; TM5. 1990. *Design of Structures to Resist the Effects of Accidental Explosions.* Department of army – TM5-1300.)

TABLE 4.2 Indicative values of heat of detonation of common explosives

EXPLOSIVES	HEAT OF DETONATION (MJ/KG)	CHARACTERISTICS	USES
TNT (trinitrotoluene)	4.10–4.55	Safe to handle; Contains insufficient oxygen for complete combustion	Popular as military and commercial explosive; Used in conjunction with ammonium nitrate–AMATOLS
C4	5.86		
RDX (cyclotrimethyl-enetrinitramine)	5.31–6.19	Thermally stable and powerful; Requires some densitization	Powerful military explosive; May be used with TNT
PETN (pentaerythritol tetranitrate)	6.69	Very powerful but too sensitive; Expensive to manufacture	Used as military explosive in combination with other explosives, e.g. with TNT-pentolites
Pentolite 50/50	5.86		
NG (nitroglycerine)	6.30	Liquid; Very sensitive explosive when solid	Used in dynamites
Nitromethane	6.40		
NC (nitrocellulose)	10.60	Sensitive and dangerous explosive	Used in mixtures with nitroglycerine–gelignite; commercial and military explosives
AN (amon./nit.)	1.59		

Sources: Vasilis Karlos, G.S. 2013. *Calculation of Blast Loads for Application to Structural Components.* Italy: European Commission – Joint research centre; Harris, R.J. 1983. *The Investigation and Control of Gas Explosions in Buildings and Heating Plant.* University of Michigan: E. & F.N. Spon in association with the British Gas Corp.

An equivalent TNT weight (*W*) is computed according to Equation 4.28 that links the weight of the chosen design explosive to the equivalent weight of TNT by utilising the ratio of the heat produced during detonation (see Table 4.2) (Vasilis 2013):

$$W = W_{exp} \frac{H_{exp}^d}{H_{TNT}^d} \tag{4.28}$$

where

W = the TNT equivalent weight (kg)
W_{exp} = the weight of the actual explosive (kg)
H_{exp}^d = the heat of detonation of the actual explosive (MJ/kg)
H_{TNT}^d = the heat of detonation of the TNT (MJ/kg)

Selection of the blast charge size 'W' is based on the perceived risk to the design structure and any structures nearby. Various factors play a role here, such as the social and economic significance of the structure, security measures that deter terrorists, and data from previous attacks on similar facilities. The minimum standoff distance 'R' is determined from the layout of a structure's surroundings and reflects the expectation of how close to the structure the design charge could explode. 'W' and 'R' are two important inputs for the scaled distance parameter Z (Farid and Longinow 2003).

The velocity of the blast wave propagation in the air (U_s) may be calculated as

$$U_s = \sqrt{\frac{6P_s + 7P_0}{7P_0}} a_0 \tag{4.29}$$

where a_0 is the sound velocity in air (m/s).

The blast wave propagation in the air produces dynamic pressure, which is calculated as

$$q_s = \frac{5P_s^2}{2(P_s + P_0)} \tag{4.30}$$

As the blast waves collide with a perpendicular surface, reflection pressure is generated, which can be evaluated using the following equation:

$$P_T = 2P_s \left[\frac{7P_0 + 4P_s}{7P_0 + P_s} \right] \tag{4.31}$$

The duration of the blast loading on a structure (t_s) can be calculated by using the diagram presented by TM5-1300 or the equation proposed by Izadifard and Maheri (2010) as

$$\log_{10}\left(\frac{t_s}{W}^{1/3} \right) = \begin{cases} 2.5\log_{10}(Z) + 0.28, & \text{for } Z \leq 1 \\ 0.31\log_{10}(Z) + 0.28, & \text{for } Z \geq 1 \end{cases} \tag{4.32}$$

Blast waves caused by gas or dust explosions differ from those of TNT or other high explosives as they generate smaller overpressures and larger impulses in the near field (Stolz, Klomfass, and Millon 2016).

Due to their highly ductile features, structural steel frames provide additional ultimate resistance for a blast event exceeding in severity the design blast.

Bolted connections, such as those using top and bottom flange angles, can sustain significant inelastic deformations and sometimes are preferred in blast-resistant design (Farid and Longinow 2003).

The book by J. M. Biggs, which is a revision of an earlier book written by several authors including J. M. Biggs, contains excellent simple methods for the design of structures subjected to blast loads produced by blast from nuclear weapons (Farid and Longinow 2003).

Brode (1955) proposed the following equation for calculating the peak static overpressure for near (when the p_s is greater than 10 bar) and for medium-to-far away (when the p_s is between 0.1 and 10 bar):

$$P_s = \begin{cases} \dfrac{6.7}{Z^3} + 1, & \text{for } P_s > 10 \text{ bar} \\[2mm] \dfrac{0.975}{Z} + \dfrac{1.455}{Z^2} + \dfrac{5.85}{Z^3} - 0.019, & \text{for } 0.1 < P_s < 10 \text{ bar} \end{cases} \tag{4.33}$$

where

Z = scaled distance
R = distance from centre of a spherical charge (m)
W = charge mass (kg of TNT)
P_s = peak static wave front overpressure

To calculate the maximum value of negative pressure (pressure below ambient pressure) in the negative phase of the blast, Brode (1955) proposed the following value for $\bar{P}$ as

For $Z > 1.6$,

$$\bar{P} = -\frac{0.35}{Z} \text{ bar} \tag{4.34}$$

where, $\bar{P}$ is the maximum value of negative pressure in the negative phase of the blast.

And the corresponding specific wave impulse at this stage, $\bar{i_s}$, is given by

$$\bar{i_s} \approx i_s \left(1 - \frac{1}{2Z}\right) \tag{4.35}$$

Mills (1987) proposed the following (Mills 1987; Corr and Tam 1998):

$$P_s = \frac{1772}{Z^3} + \frac{114}{Z^2} + \frac{108}{Z} - 0.019\,\text{kPa} \tag{4.36}$$

where, Z = scaled distance $(\text{m/kg}^{1/3})$.

Newmark also proposed an equation for calculating peak overpressure for ground surface blast as

$$P_{so} = 6784\frac{W}{R^3} + 93\sqrt{\frac{W}{R^3}} \tag{4.37}$$

where

P_{so} = peak overpressure (bars)
W = the charge mass in metric tonnes (= 1000 kg) of TNT
R = the distance of the surface from the centre of a spherical explosion (m)

4.2.5 Transportation Loads

Any unit should be able to take the loading from

- Roll plus heave
- Pitch plus heave
- 10-year, 1-minute wind
- Hog or sag from wave whilst being lowered through the splash zone

4.2.6 Installation Loads

Installation or construction loads depend on the mode of installation and prevailing environmental conditions. Typical loads during installation include the following:

- Fabrication
- Loadout
- Transportation
- Installation

4.2.7 Typical Load Conditions

The following lists typical load conditions that may be required to be considered:

- Dead load, frame
- Dead load, fixtures
- Dead load, equipment

- Wind loads (N–S)
- Wind loads (E–W)
- Operational loads
- Maintenance loads
- Dropped objects
- Trawl board impact
- Hydrodynamic loads
- Hydrostatic collapse
- Lift in air
- Lift in water
- Lift through splash zone
- Temporary stability
- Installed loads
- Pilling loads
- Installed pile and structure interaction
- Sea transportation

4.2.8 Load Factors

Often a load case is required to be factored for a particular load combination. For example, in a lift analysis, it may be required to use a load factor of 2.0.

4.2.9 Load Combination

The aim of deriving load combinations is to establish, for analysis, the various combined loadings a structure will have to withstand.

Load combination is the adding of one or more factored or unfactored load conditions (see Table 4.3). All loadings should be defined individually

TABLE 4.3   Load combination examples

LOAD CASE NO.	DESCRIPTION	LOAD FACTORS						
		D +	L +	I +	E +	H +	A +	B
1	Lift	2.0	0.0	0.0	0.0	0.0	0.0	0.0
2	Transportation	1.0	0.0	1.0	0.0	0.0	0.0	0.0
3	On bottom normal	1.0	1.0	0.0	1.0	1.0	0.0	1.0
4	On bottom extreme	1.0	1.0	0.0	1.0	1.0	1.0	1.0

Source: SJR. 1994. Subsea Structure Design. London: JP Kenny Group.
Note: A = Accidental loads, B = Buoyancy loads, D = Dead loads, E = Environmental loads, H = Hydrostatic loads, I = Transportation inertia loads, L = Live loads.

and then subsequently combined to form design load combinations (SJR 1994; Nallayarasu 2013). The offshore platform should be designed to satisfy all code requirements whilst under the following conditions:

a. Loadout
b. Transportation
c. Installation
d. Operation/in service

The following elements of each of these phases determine the load conditions and combinations:

Loadout:

- Lifting
- Skidding
- Jacking up and rolling
- Launching

Transportation:

- Transportation as vessel deck cargo
- Transportation by flotation and towing

Installation:

- Lifting and lowering
- Temporary stability and support conditions
- Pilling effects
- Tie-in spool or jumper installation
- Equipment installation

In service:

- Equipment loads
- Pipeline or piping system forces, moments and displacements
- Fatigue
- Environmental loading
- Live and maintenance loading
- Damage

The load combination recommended for use with allowable stress design procedures are

a. During normal operations

Load combination = Dead loads + Operating environmental loads
+ Maximum live loads + Accidental loads

Or

Load combination = Dead loads + Operating environmental loads
+ Minimum live loads + Accidental loads

b. Extreme conditions

Load combination = Dead loads + Extreme environmental loads
+ Maximum live loads + Accidental loads

Or

Load combination = Dead loads + Extreme environmental loads
+ Minimum live loads + Accidental load

Earthquake loads are to be imposed as a separate environmental load. Earthquake load should not be combined with waves, wind, and so forth.

4.2.9.1 Load Combination for Accidental Fire Situations

The applied load is obtained by considering the accidental combination of the mechanical actions, such as dead load, live load, wind (only for bracing) and snow.

Due to the low probability that both fire and extreme severity of external actions occur at the same time, only the following accidental combinations are considered (ESDEP 2017).

$$1.0G_k + y_1Q_{k,1} + Sy_{2,i}Q_{k,i} \tag{4.38}$$

where

G_K = the characteristic value of permanent actions (permanent or dead load)
$Q_{K,1}$ = the characteristic value of the main variable actions
$Q_{K,i}$ = the characteristic value of other variable actions
y_1 = the frequent value of the main variable actions
$y_{2,i}$ = the average of the other variable actions

Generally, in fire: $y_1 = 0.5$ and $y_{2,i} = 0$.

Apart from bracings, $Q_{K,1}$ and $Q_{K,2}$ generally correspond to imposed loads and snow loads.

4.3 DESIGN OF TUBULAR MEMBERS

Most offshore platforms used for drilling and production of oil and gas are constructed of tubular members welded together to form jackets or templates on which drilling and production modules are supported.

The load carrying capacity of a cylindrical or tubular member is dependent on its

- Length
- Diameter
- Wall thickness
- Material properties, (elastic modulus [E], specified minimum yield strength [SMYS])
- End and intermediate restraints
- The manner in which the loads are applied

The stresses induced in a member may be

- Bending
- Compression
- Tensile
- Shear
- Torsion
- Hoop

Where applicable, these stresses may be combined to arrive at a combined stress (e.g. Von Mises).

In the design of a member, the ultimate load-carrying capacity and the behaviour/deflection under working load conditions are of interest. At present, it is usual to analyse the elastic behaviour of a structure and approximate the load-carrying capacity of its members by applying factored yield stress to the design.

The design of a member will need to allow for all, or some, of the following:

- Imposed loads and/or self-weight
- Corrosion allowance
- Accidental damage

- Energy absorption
- Environment (e.g. hydrostatic and hydrodynamic forces)
- Fatigue

The allowable stress for working load conditions, in part, is dependent on the nature of the loadings and the purpose of the member. A member may be required to permanently, temporarily or only once carry a load. Structures can be designed to 'allowable stresses' or 'limit state' methods.

a. Allowable stress: Permanent and temporary structures
b. Ultimate limit state: Accidental damage

For the design of the tubular members of the offshore structures for accidental fire condition, there is a need to use the 'Ultimate limit state methods' for design.

This section details the design of tubular or cylindrical members. The stress calculations are based on the requirements of API RP 2A and Chapter 13 of ISO 19902.

4.3.1 Determination of Component of Stresses

The calculations in this section apply to unstiffened and ring-stiffened cylindrical tubulars having a thickness $t \geq 6$ mm, a diameter to thickness ratio $D/t \leq 120$ and material should fulfil Clause 19 of ISO 19902, with yield strengths less than 500 MPa. The ratio of yield strength as used to ultimate tensile strength should not exceed 0.90.

Divide the member into sections and calculate the stress in the tubular members as shown in Equations 4.39 to 4.97 below.

4.3.1.1 Axial Tension

Tubular or cylindrical members subjected to axial tensile forces should be designed to satisfy Equation 4.39.

$$\sigma_t \leq \frac{f_t}{\gamma_{R,t}} \tag{4.39}$$

where

σ_t = axial tensile stress due to forces from factored actions
f_t = the representative axial tensile strength
f_y = the representative yield strength, in stress units
$\gamma_{R,t}$ = the partial resistance factor for axial tensile strength, $\gamma_{R,t} = 1.05$

The utilisation of a member, U_m, under axial tension is calculated as

$$U_m = \frac{\sigma_t}{(f_t/\gamma_{R,t})} \tag{4.40}$$

4.3.1.2 Axial Compression

Tubular members subjected to axial compressive forces shall be designed to satisfy Equation 4.41:

$$\sigma_c \leq \frac{f_c}{\gamma_{R,c}} \tag{4.41}$$

where
   σ_c = axial compressive stress due to forces from factored actions
   f_c = the representative axial compressive strength, in stress units
   $\gamma_{R,c}$ = the partial resistance factor for axial compressive strength, $\gamma_{R,c} = 1.18$

The utilisation of a member, U_m, under axial compression is calculated as

$$U_m = \frac{\sigma_c}{(f_c/\gamma_{R,c})} \tag{4.42}$$

4.3.1.2.1 Column Buckling
When there is no hydrostatic pressure, the representative axial compressive strength calculated in Equation 4.43 for tubular members shall be the smaller of the in-plane and the out-of-plane buckling strengths determined from Equations 4.43 and 4.44:
   For $\lambda \leq 1.34$,

$$f_c = (1.0 - 0.278\lambda^2)f_{yc} \tag{4.43}$$

   For $\lambda > 1.34$,

$$f_c = \frac{0.9}{\lambda^2} f_{yc} \tag{4.44}$$

where
   f_c = the representative axial compressive strength, in stress units
   f_{yc} = the representative local buckling strength, in stress unit
   λ = is the column slenderness parameter

$$\lambda = \sqrt{\frac{f_{yc}}{f_e}} = \frac{kL}{\pi r}\sqrt{\frac{f_{yc}}{E}} \qquad (4.45)$$

f_e = the smaller of the Euler buckling strengths in the y- and z-directions, in stress units
E = Young's modulus of elasticity
k = the effective length factor in the y- or z-direction selected so that kL is the larger of the values in the y- and z-directions
L = the unbraced length in the y- or z-direction
R = the radius of gyration, $r = I/A$
I = the moment of inertia of the cross-section
A = the cross-sectional area

4.3.1.2.2 Local Buckling

Global buckling is taken into account by means of the slenderness ratio or the multiplication factor called effective length factor (L).

The failure of offshore structure under various buckling modes should be investigated in terms of elastic buckling stress and inelastic buckling stress (ISO 2007).

The representative local buckling strength, f_{yc}, is calculated as follows.

For $(f_y/f_{xe}) \leq 0.170$,

$$f_{yc} = f_y \qquad (4.46)$$

For $(f_y/f_{xe}) > 0.170$,

$$f_{yc} = \left(1.047 - 0.274\frac{f_y}{f_{xe}}\right)f_y \qquad (4.47)$$

$$f_{xe} = 2C_x Et/D \qquad (4.48)$$

where

f_y = the representative yield strength, in stress units
f_{xe} = the representative elastic local buckling strength, in stress units
C_x = the elastic critical buckling coefficient
E = Young's modulus of elasticity
D = the outside diameter of the member
t = the wall thickness of the member

The theoretical value of C_x for an ideal tubular is 0.6. However, a reduced value of $C_x = 0.3$ should be used in Equation 4.48 to account for the effect of initial geometric imperfections within the tolerance limits given in Clause 21 of ISO 19902. A reduced value of $C_x = 0.3$ is implicit in the value of f_{xe} used in Equations 4.46 and 4.47.

4.3.1.3 Bending

Tubular/cylindrical members subjected to bending moments shall be designed to fulfil Equation 4.49.

$$\sigma_b = \frac{M}{Z_e} \le \frac{f_b}{\gamma_{R,b}} \qquad (4.49)$$

where

σ_b = the bending stress due to forces from factored actions; when $M > My$, σ_b is to be considered as an equivalent elastic bending stress, $\sigma_b = (M/Z_e)$
f_b = the representative bending strength, in stress units
$\gamma_{R,b}$ = the partial resistance factor for bending strength, $\gamma_{R,b} = 1.05$
M = the bending moment due to factored actions
My = the elastic yield moment
Z_e = elastic plastic modulus

$$Z_e = \frac{\pi}{64} \frac{(D^4 - (D - 2t)^4)}{(D/2)}$$

The utilisation of a member, U_m, under bending moments is calculated using Equation 4.50.

$$U_m = \frac{\sigma_b}{f_b/\gamma_{R,b}} = \frac{M/Z_e}{f_b/\gamma_{R,b}} \qquad (4.50)$$

The representative bending strength for tubular members shall be calculated using Equations 4.51 to 4.53.

For $f_y D/Et \le 0.0517$,

$$f_b = \left(\frac{Z_p}{Z_e}\right) f_y \qquad (4.51)$$

For $0.0517 < f_y D/Et \le 0.1034$,

$$f_b = \left[1.13 - 2.58\left(\frac{f_y D}{Et}\right)\right]\left(\frac{Z_p}{Z_e}\right) f_y \qquad (4.52)$$

For $0.1034 < f_y D/Et \leq 120 f_y/E$,

$$f_b = \left[0.94 - 0.76\left(\frac{f_y D}{Et}\right)\right]\left(\frac{Z_p}{Z_e}\right)f_y \tag{4.53}$$

where

f_y = the representative yield strength, in stress units
D = the outside diameter of the member
t = the wall thickness of the member
Z_p = the plastic section modulus, $Z_p = (1/6)(D^3 - (D - 2t)^3)$

4.3.1.4 Shear

4.3.1.4.1 Beam Shear

Tubular members subjected to beam shear should be designed to satisfy Equation 4.54.

$$\tau_b = \frac{2V}{A} \leq \frac{f_v}{\gamma_{R,V}} \tag{4.54}$$

where

τ_b = the maximum beam shear stress due to forces from factored actions
f_v = the representative shear strength, in stress units, $f_v = f_y/\sqrt{3}$
$\gamma_{R,V}$ = the partial resistance factor for shear strength, $\gamma_{R,V} = 1.05$
V = the beam shear due to factored actions, in force units
A = the cross-sectional area

The utilisation of a member, U_m, under beam shear shall be calculated using Equation 4.55.

$$U_m = \frac{\tau_b}{(f_v/\gamma_{R,V})} = \frac{2V/A}{(f_v/\gamma_{R,V})} \tag{4.55}$$

4.3.1.4.2 Torsional Shear

$$\tau_t = \frac{M_{v,t}\,D}{2I_p} \leq \frac{f_v}{\gamma_{R,V}} \tag{4.56}$$

where

τ_t = torsional shear stress due to forces from factored actions
$M_{v,t}$ = the torsional moment due to factored actions
I_p = the polar moment of inertia, $I_p = \pi/32[D^4 - (D - 2t)^4]$

The utilisation of a member, U_m, under torsional shear shall be calculated using Equation 4.57.

$$U_m = \frac{\tau_t}{(f_v/\gamma_{R,V})} = \frac{(M_{v,t}, D/2I_p)}{(f_v/\gamma_{R,V})} \tag{4.57}$$

All other parameters are as defined above. Refer to Clause 13.2.6 of ISO 19902 for the calculation of hydrostatic pressure.

4.3.1.5 Hoop Buckling

Tubular or cylindrical members subjected to external pressure should be designed to satisfy Equation 4.58.

$$\sigma_h = \frac{PD}{2t} \leq \frac{f_h}{\gamma_{R,h}} \tag{4.58}$$

where

σ_h = the hoop stress due to the forces from factored hydrostatic pressure
P = the factored hydrostatic pressure
D = the outside diameter of the member
f_h = the representative hoop buckling strength, in stress units (see Equations 4.59 through 4.61)
$\gamma_{R,h}$ = the partial resistance factor for hoop buckling strength, $\gamma_{R,h} = 1.25$

For tubular members satisfying the out-of-roundness tolerances, f_h, should be calculated as follows.
For $f_{he} > 2.44f_y$,

$$f_h = f_y \tag{4.59}$$

For $0.55f_y < f_{he} \leq 2.44f_y$,

$$f_h = 0.7\left(\frac{f_{he}}{f_y}\right)^{0.4} f_y \leq f_y \tag{4.60}$$

For $f_{he} \leq 0.55f_y$,

$$f_h = f_{he} \tag{4.61}$$

where

f_y = the representative yield strength, in stress units and
f_{he} = the representative elastic critical hoop buckling strength, in stress units.

$$f_{he} = 2C_h Et/D \tag{4.62}$$

where the elastic critical hoop buckling coefficient, C_h, is as follows:
For $\mu \geq 1.6D/t$,

$$C_h = \frac{0.44t}{D} \tag{4.63}$$

For $0.825D/t \leq \mu < 1.6D/t$,

$$C_h = \frac{0.44t}{D} + 0.21\left(\frac{D}{t}\right)^3 \mu^4 \tag{4.64}$$

For $1.5 \leq \mu < 0.825D/t$,

$$C_h = \frac{0.737}{(\mu - 0.579)} \tag{4.65}$$

For $\mu < 1.5$,

$$C_h = 0.80 \tag{4.66}$$

where μ is a geometric parameter and

$$\mu = \frac{L_T}{D}\sqrt{\frac{2D}{t}} \tag{4.67}$$

where L_T is the length of tubular between stiffening rings, diaphragms, or end connections.

For tubular members exceeding the out-of-roundness tolerances, the utilisation of a member, U_m, under external pressure should be calculated as

$$U_m = \frac{\sigma_h}{f_h/\gamma_{R,h}} = \frac{(PD/2t)}{f_h/\gamma_{R,h}} \tag{4.68}$$

4.3.1.6 Ring Stiffener Design

For $\mu \geq 1.6\,D/t$, the elastic critical hoop buckling stress is roughly equal to that of a long unstiffened tubular. Therefore, to be operational, stiffening rings, if required, should be spaced such that

$$L_r < 1.6\sqrt{\frac{D^3}{2t}} \tag{4.69}$$

Use Equations 4.70 and 4.71 to calculate the circumferential stiffening ring size as applicable, on condition that the yield strength of the ring stiffener is less than that of the member and that this smaller value of yield strength is used instead of f_j in Equations 4.59 to 4.61.

For internal rings,

$$I_c = f_{he} \frac{tL_r D^2}{8E} \tag{4.70}$$

For external rings,

$$I_c = f_{he} \frac{tL_r D_r^2}{8E} \tag{4.71}$$

where
I_c = the required moment of inertia for the composite ring section
L_r = the ring spacing
D = the outside diameter of the member
D_r = the diameter of the centroid of the composite ring section
E = young modulus of elasticity

All other parameters are as defined in previous sections above. If Equations 4.72 and 4.73 are satisfied, then local buckling of ring stiffeners with flanges may be excluded.

$$\frac{h}{t_w} \leq 1.1 \sqrt{\frac{E}{f_{y,r}}} \tag{4.72}$$

and

$$\frac{b}{t_f} \leq 0.6 \sqrt{\frac{E}{f_{y,r}}} \tag{4.73}$$

where
h = the web height
t_w = the web thickness
b = half the flange width of T stiffeners or the full flange width for angle stiffeners
t_f = flange thickness
$f_{y,r}$ = the representative yield strength of the ring stiffeners, in stress units.

For ring stiffeners without flanges, local buckling may be overcome if Equation 4.74 is satisfied.

$$\frac{h}{t_w} \le 0.6\sqrt{\frac{E}{f_{y,r}}} \tag{4.74}$$

Ring stiffeners, together with their components and with internal or external rings, should have a minimum thickness of 10 mm.

4.3.1.7 Tubular Members Subjected to Combined Forces without Hydrostatic Pressure

4.3.1.7.1 Axial Tension and Bending
Tubular members subjected to a combined axial tension and bending forces should satisfy Equation 4.75 at all cross-sections along their length.

$$\frac{\gamma_{R,t}\sigma_t}{f_t} + \frac{\gamma_{R,b}\sqrt{\sigma_{b,y}^2 + \sigma_{b,z}^2}}{f_b} \le 1.0 \tag{4.75}$$

where
$\sigma_{b,y}$ = the bending stress about the member y-axis (in-plane) due to forces from factored actions and
$\sigma_{b,z}$ = the bending stress about the member z-axis (out-of-plane) due to forces from factored actions.

All other parameters are as defined in previous equations above.
The utilisation of a member, U_m, under combined axial tension and bending should be calculated from Equation 4.76.

$$U_m = \frac{\gamma_{R,t}\sigma_t}{f_t} + \frac{\gamma_{R,b}\sqrt{\sigma_{b,y}^2 + \sigma_{b,z}^2}}{f_b} \tag{4.76}$$

4.3.1.7.2 Axial Compression and Bending
Tubular members subjected to combined axial compression and bending forces shall be designed to satisfy the following conditions at all cross-sections along their length.

$$\frac{\gamma_{R,c}\sigma_c}{f_c} + \frac{\gamma_{R,b}}{f_b}\left[\left(\frac{C_{m,y}\sigma_{b,y}}{1 - \sigma_c/f_{e,y}}\right)^2 + \left(\frac{C_{m,z}\sigma_{b,z}}{1 - \sigma_c/f_{e,z}}\right)^2\right]^{0.5} \le 1.0 \tag{4.77}$$

and

$$\frac{\gamma_{R,c}\sigma_c}{f_{yc}} + \frac{\gamma_{R,b}\sqrt{\sigma_{b,y}^2 + \sigma_{b,z}^2}}{f_h} \leq 1.0 \tag{4.78}$$

where

$C_{m,y}$, $C_{m,z}$ are the moment reduction factors corresponding to the member y- and z-axes, respectively and

$f_{e,y}$, $f_{e,z}$ are the Euler buckling strengths corresponding to the member y- and z-axes, respectively, in stress units.

$$f_{e,y} = \frac{\pi^2 E}{(K_y L_y / r)^2} \tag{4.79}$$

$$f_{e,z} = \frac{\pi^2 E}{(K_z L_z / r)^2} \tag{4.80}$$

where

K_y, K_z = the effective length factors for the y- and z-directions, respectively and

L_y, L_z = the unbraced lengths in the y- and z-directions, respectively.

The utilisation of a member, U_m, under axial compression and bending shall be the larger value calculated from Equations 4.81 and 4.82.

$$U_m = \frac{\gamma_{R,c}\sigma_c}{f_c} + \frac{\gamma_{R,b}}{f_b}\left[\left(\frac{C_{m,y}\sigma_{b,y}}{1 - \sigma_c/f_{e,y}}\right)^2 + \left(\frac{C_{m,z}\sigma_{b,z}}{1 - \sigma_c/f_{e,z}}\right)^2\right]^{0.5} \tag{4.81}$$

$$U_m = \frac{\gamma_{R,c}\sigma_c}{f_{yc}} + \frac{\gamma_{R,b}\sqrt{\sigma_{b,y}^2 + \sigma_{b,z}^2}}{f_b} \tag{4.82}$$

4.3.1.8 Tubular Members Subjected to Combined Forces with Hydrostatic Pressure

A tubular member below the waterline is subjected to hydrostatic pressure unless it has been flooded.

For tubular members subjected to hydrostatic pressure, four checks are compulsory:

a. Check for hoop buckling under hydrostatic pressure alone.
b. Check for tensile yielding when the combination of action effects, including those due to capped-end forces, results in tension in the member.

c. Check for compression yielding and local buckling when the combination of action effects, including those due to capped-end forces, results in compression in the member.

d. Check for column buckling when the action effects, excluding those due to capped-end forces, result in compression in the member.

Refer to Clause 13.4.1 to Clause 13.4.3 of ISO 19902 for equations concerning the four mandatory checks stated above.

4.3.1.8.1 Axial Tension, Bending and Hydrostatic Pressure

Tubular members subjected to combined axial tension, bending and hydrostatic pressure shall be designed to satisfy Equation 4.83 at all cross-sections along their length.

$$\frac{\gamma_{R,t}\sigma_{t,c}}{f_{t,h}} + \frac{\gamma_{R,b}\sqrt{\sigma_{b,y}^2 + \sigma_{b,z}^2}}{f_{b,h}} \leq 1.0 \tag{4.83}$$

where

$f_{t,h}$ = the representative axial tensile strength in the presence of external hydrostatic pressure, in stress units

$$f_{t,h} = f_y\left(\sqrt{1+0.09B^2 - B^{2\eta}} - 0.3B\right) \tag{4.84}$$

$f_{b,h}$ = the representative bending strength in the presence of external hydrostatic pressure, in stress units

$$f_{b,h} = f_b\left(\sqrt{1+0.09B^2 - B^{2\eta}} - 0.3B\right) \tag{4.85}$$

And for $B \leq 1.0$

$$B = \frac{\gamma_{R,h}\sigma_h}{f_h} \tag{4.86}$$

$$\eta = 5 - 4\frac{f_h}{f_y} \tag{4.87}$$

The utilisation of a member, U_m, under axial tension, bending and hydrostatic pressure shall be calculated from Equation 4.88.

$$U_m = \frac{\gamma_{R,t}\sigma_{t,c}}{f_{t,h}} + \frac{\gamma_{R,b}\sqrt{\sigma_{b,y}^2 + \sigma_{b,z}^2}}{f_{b,h}} \tag{4.88}$$

4.3.1.8.2 Axial Compression, Bending and Hydrostatic Pressure

Tubular members subjected to combined axial compression, bending and hydrostatic pressure shall be designed to satisfy Equations 4.89 and 4.90 at all cross-sections along their length.

$$\frac{\gamma_{R,c}\sigma_{c,c}}{f_{yc}} + \frac{\gamma_{R,b}\sqrt{\sigma_{b,y}^2 + \sigma_{b,z}^2}}{f_{b,h}} \leq 1.0 \tag{4.89}$$

$$\frac{\gamma_{R,c}\sigma_c}{f_{c,h}} + \frac{\gamma_{R,b}}{f_{b,h}}\left[\left(\frac{C_{m,y}\sigma_{b,y}}{1-\sigma_c/f_{e,y}}\right)^2 + \left(\frac{C_{m,z}\sigma_{b,z}}{1-\sigma_c/f_{e,z}}\right)^2\right]^{0.5} \leq 1.0 \tag{4.90}$$

where

$f_{c,h}$ = the representative axial compressive strength in the presence of external hydrostatic pressure, in stress units

For $\lambda \leq 1.34\sqrt{\left(1-\dfrac{2\sigma_q}{f_{yc}}\right)^{-1}}$

$$f_{c,h} = 1/2 f_{yc}\left[\frac{(1.0-0.278\lambda^2)-\dfrac{2\sigma_q}{f_{yc}}}{+\sqrt{(1.0-0.278\lambda^2)^2 + 1.12\lambda^2\dfrac{\sigma_q}{f_{yc}}}}\right] \tag{4.91}$$

For $\lambda > 1.34\sqrt{\left(1-\dfrac{2\sigma_q}{f_{yc}}\right)^{-1}}$

$$f_{c,h} = \frac{0.9}{\lambda^2}f_{yc} \tag{4.92}$$

If the maximum combined compressive stress, $\sigma_x = \sigma_b + \sigma_{c,c}$, and the representative elastic local buckling strength, f_{xe}, exceed the limits given in Equation 4.93, then Equation 4.94 shall also be satisfied:

$$\sigma_x > 0.5\frac{f_{he}}{\gamma_{R,h}} \text{ and } \frac{f_{xe}}{\gamma_{R,c}} > 0.5\frac{f_{he}}{\gamma_{R,h}} \tag{4.93}$$

$$\left[\frac{\sigma_x - (0.5(f_{he}/\gamma_{R,h}))}{(f_{xe}/\gamma_{R,c})-(0.5(f_{he}/\gamma_{R,h}))}\right] + \left(\frac{\gamma_{R,h}\sigma_h}{f_{he}}\right)^2 \leq 1.0 \tag{4.94}$$

where

f_{he} = the representative elastic critical hoop buckling strength defined in Equation 4.62

f_{xe} = the representative elastic local buckling strength defined in Equation 4.48.

The utilisation of a member, U_m, under axial compression, bending and hydrostatic pressure shall be the largest value calculated from Equations 4.95 to 4.97.

When Equation 4.89 applies,

$$U_m = \frac{\gamma_{R,c}\sigma_{c,c}}{f_{yc}} + \frac{\gamma_{R,b}\sqrt{\sigma_{b,y}^2 + \sigma_{b,z}^2}}{f_{b,h}} \tag{4.95}$$

When Equation 4.90 applies,

$$U_m = \frac{\gamma_{R,c}\sigma_c}{f_{c,h}} + \frac{\gamma_{R,b}}{f_{b,h}}\left[\left(\frac{C_{m,y}\sigma_{b,y}}{1 - \sigma_c/f_{e,y}}\right)^2 + \left(\frac{C_{m,z}\sigma_{b,z}}{1 - \sigma_c/f_{e,z}}\right)^2\right]^{0.5} \tag{4.96}$$

When Equation 4.94 applies,

$$U_m = \left[\frac{\sigma_x - (0.5(f_{he}/\gamma_{R,h}))}{(f_{xe}/\gamma_{R,c}) - (0.5(f_{he}/\gamma_{R,h}))}\right] + \left(\frac{\gamma_{R,h}\sigma_h}{f_{he}}\right)^2 \tag{4.97}$$

4.3.2 Slenderness Ratio

Slenderness ratio is the ratio of the effective length of a column (*L*) and the least radius of gyration (*r*) about the axis under consideration. It is given by the symbol 'λ' (lambda). As slenderness ratio increases, permissible stress or critical stress reduces. Consequently, load carrying capacity also reduces.

The slenderness ratio is calculated as

$$\text{Slenderness ratio } (\lambda) = \frac{\text{Effective length}}{\text{Least radius of gyration}} \tag{4.98}$$

$$\lambda = \frac{kL}{r} \tag{4.99}$$

where

kL = effective member length depending on end-restraint conditions,

L = actual member length,

k = the effective length factor or effective length coefficient and
r = cross-sectional radius of gyration.

$$r = \left(\frac{I}{A}\right)^{1/2} \qquad (4.100)$$

Depending on the structure end conditions or boundary conditions, the value of 'k' may vary. The effective length factor, k, can be obtained from Table 13.5-1 of ISO 19902.

In the seismic zone, the slenderness ratio (kL/r) of primary bracing in vertical frames shall be limited to no more than 80 (with the corresponding column slenderness parameter λ not exceeding $(80/\pi) \times \sqrt{f_{yc}/E}$, and $f_y(D/E) \times t \le 0.069$.

For values of $30 < kL/r > 100$, the slenderness ratio is considered to be within the middle range.

Most designers aim to maintain slenderness ratios between 60 and 90; within this range, the member strength depends on the tangent modulus of the material and on end-restraint design (Bea 1991).

In a fire situation, ductile collapse of the structure may be accelerated in designs with high slenderness ratios, even though such designs may employ a higher degree of redundancy. Lower slenderness ratios encourage high D/t ratios for tubular members that may compound local buckling problems.

In seismic zones, the slenderness ratio of primary diagonal bracing in vertical frames is limited to a maximum of 80, and the D/t ratio is restricted to 1900/Fy (ksi).

For a brace, it is advisable to limit the slenderness ratio to 90 in order to avoid Euler and local buckling (James 1988; Bea 1991). For optimum deck plate design, plate slenderness in the range of 70–100 is recommended for design. At this range, the change of response to blast is more predictable (Ali 2007).

4.3.3 Wall Thickness Modulus

The wall thickness modulus is used to classify a tubular section as thin or thick wall members and is a measure of buckling resistance.

$$\frac{D}{t} \qquad (4.101)$$

where

D = diameter of tubular member
t = wall thickness of the tube

It is practical to keep the D/t ratios between 30 and 60. Tubular members with D/t ratios less than or equal to 60 are normally not subjected to local buckling from axial compression and can be designed on the basis of material failure. D/t greater than 60 can present local buckling problems. By using compact sections for offshore designs, local buckling of tubular compression members can be avoided. Sizes of braces may be reduced to lower the drag force and loadings on the structure (Bea 1991).

Hence, thinner-wall tubular members are practically sensitive to failure from local buckling, either due to fabrication defects or thermal impact.

Members with D/t ratios below 25 are considered thick-walled and will not float. Therefore, their use offshore has been limited to date. On the other hand, members with low D/t ratios have much greater inherent thermal mass and fire endurance than thinner-wall members, and may find greater use in the future for above-waterline applications for reasons of their increased thermal robustness (Bea 1991).

4.3.4 Accidental Impact Loads

In accidental load cases, stress levels in excess of yield are of concern. For the purposes of design, the allowable stress may be predictably taken as the yield stress. However, for analysis of a damaged member, consideration of the actual stresses occurring at the time of damage should be made.

The allowable design stress, F_p, is calculated as

$$F_p = 1.0 \times F_y = F_y \tag{4.102}$$

The allowable stress may exceed yield in cases where a permanent deflection (set) of the member is permissible. The stress in this case may be increased by a percentage equal to the positive tolerance of F_y (SMYS), which may be obtained from the appropriate material specification. In the absence of these data, a permissible overstress of 9% may be used (SJR 1994).

4.3.5 Temporary and Transient Loads

Examples of temporary structures and members include: pile guides, docking piles, pins, buckets, bumpers, and so on. Typical temporary and transient loads are from sea transportation and wind respectively. Allowable stress may be increased by multiplying it with a factor; API RP 2A states that for temporary and transient load conditions, one-third increase is allowable (SJR 1994).

Allowable working stress in tension is calculated as

$$F_t = 0.6 F_y \tag{4.103}$$

Allowable wind stress in tension is calculated as

$$F_t = \frac{4}{3} \times 0.6 F_y \qquad\qquad (4.104)$$

4.3.6 Collapse of Members Subjected to Hydrostatic Pressure

A hollow, empty sealed member subjected to a large enough hydrostatic pressure will collapse. This failure of members in offshore structures must be considered and designed against. Such failure is, in part, related to the geometry and dimensions of the cross-section of the member.

Circular cross-sectional members are structurally more efficient than other shapes when subjected to hydrostatic pressures and should be used wherever possible. Rectangular hollow sections should be avoided when the member is required to resist hydrostatic pressure. Rectangular hollow sections may be used for secondary structures if allowed to free flood, particularly where high bending moments are imposed (SJR 1994).

Hydrostatic collapse can be avoided by

- Sufficient wall thickness
- Reinforcing the walls with internal and/or external stiffeners
- Filling the member with water or grout

The hydrostatic pressure used in the calculation is derived from a combination of the still water depth above the member and wave height. In shallow waters, the wave height can add significantly to the pressures.

4.3.7 Conical Transition between Cylindrical Sections

These are used to join cylindrical members of common axis but of different diameters. A cone must be designed for the combination of axial, bending and hydrostatic forces and fatigue conditions occurring at the member's ends and within its length.

Transition cones may be strengthened by the provision of internal and/or external stiffeners or grout. It is preferable to use a transition cone of similar thickness to one of the connected tubes (SJR 1994).

Connectors should not be fitted to transition cones like padeyes, braces, and so on. Conical transitions may be checked according to Clause 13.6 of ISO 19902 or NORSOK N-004.

4.3.8 Strength of Damaged Cylindrical Members

To assess the ultimate capacity of a damaged member, these assumptions should be made:

- The loading is a point load
- The energy absorbed by the impinging object is nil
- The member supports do not deflect/sink

The amount of energy a member can absorb is partly dependent on its flexural and axial restraints. Therefore, in order to design conservatively, a member should be considered to take all the energy of impact (SJR 1994).

Clause 13.6 of ISO 19902:2007 provides the detailed procedure for calculating the design stresses in a conical transition.

4.4 JOINT DESIGN

Tubular joints occur at the structure node points (see Figure 4.7). Joints are categorised as simple, overlapping or congested joints.

Simple joints are those in which the brace members do not make contact with each other and do not have gussets, diaphragms, stiffeners, and so on, at the joints. The minimum recommended clear distance (gap) between braces at a simple joint is 50 mm.

Overlapping joints are those in which the brace members make contact with each other and are directly connected to each other at the joint. A brace

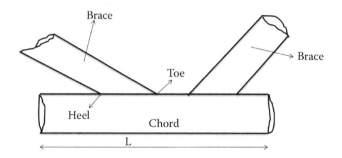

FIGURE 4.7   Simple tubular joint.

overlap should be proportioned to take 50% of the bracing component normal to the chord (SJR 1994; API 2000). Where the braces are substantially different in size or loading, the heavier brace should pass through. Congested joints occur when braces in different planes tend to overlap. When primary brace walls are thicker than secondary braces, the primary brace should pass through. A congested joint can be simplified by adjusting the brace eccentricities or by providing a larger diameter chord member at the joint (SJR 1994; API 2000). The brace wall thickness should never exceed the chord wall thickness. When the chord diameter is substantially larger than that of the brace, the effect of brace eccentricities may be ignored unless the eccentricity exceeds one quarter of the diameter of the chord.

If an increase in the wall thickness of the chord at the joint is required, it should extend past the outside of the bracing for a minimum of 300 mm or one-quarter of the chord diameter, whichever is greater.

Where increase in wall thickness of brace or special steel is used for braces in the tubular joint area, it should extend for a minimum of one brace diameter or 600 mm, whichever is greater, from the joint (API 2000).

API RP 2A requires that the connections at the ends of tension and compression members should develop the strength required by design loads, but not less than 50% of the effective strength of the member (API 2000). This need not apply to secondary attachments to the primary structure.

Tubular joints to members of varying diameter should be avoided. An example is a transition cone.

Where brace and chord members are of similar diameter, or where they incorporate stiffeners, gussets, and so forth, the joint can be analysed by cutting it into sections and carrying out ring analysis on the sections (API 2000).

Ring analysis may be required to assess stress levels where there is a tendency for the chord of a joint to be crushed, for example, T or X members. For this analysis, three times the chord diameter may be taken as the effective length of the chord. The chord can be strengthened by increasing the wall thickness locally by providing internal stiffeners or by increasing the grade of chord material. Tubular joints should be checked against punching shear.

4.4.1 Loading Conditions

- Punching shear due to the portion of the loads from the bracings resisted by the chord
- The component of the brace loads that tend to crush the chord
- Moment on the chord due to brace eccentricities

- Member loads and moments at the joint
- Fatigue

4.4.2 Punching Shear Analysis

The punching shear analysis is also known as 'joint can analysis'.

4.4.2.1 Acting Punching Shear

The acting punching shear is the shear stress developed in the chord by the brace load. With reference to API RP 2A, acting punching shear is calculated as

$$V_p = \tau f \sin\theta \qquad (4.105)$$

where
V_p = the acting punching shear
f = the nominal axial, in-plane bending or out-of-plane bending stress in the brace

The acting punching shear is calculated separately for each of these stress components (ESDEP 2017).

4.4.2.2 Chord Design Factors

With reference to API RP 2A-WSD, Chord design factor is calculated as

$$Q_f = 1.0 - \lambda\gamma A^2 \qquad (4.106)$$

where
$\lambda = 0.030$ brace axial stress
$\lambda = 0.045$ brace in-bending
$\lambda = 0.021$ brace out-bending

$$A = \frac{\sqrt{f_{axc}^2 + f_{ipc}^2 + f_{opc}^2}}{\mu f_{yc}} \qquad (4.107)$$

μ = 0.6 ordinary load case
μ = 0.8 extreme load case
μ = 1.0 earthquake load case
$Q_f = 1.0$ if all extreme fibre stresses in the chord are tensile

4.4.2.3 Geometry and Load Factors

With reference to API RP 2A-WSD, geometry and load factor is calculated as
If $\beta > 0.6$
Then

$$Q_\beta = \frac{0.3}{\beta(1 - 0.833\beta)} \tag{4.108}$$

Else $Q_\beta > 1.0$
For K joints
If $\gamma > 20$
Then

$$Q_g = 1.8 - \frac{4g}{D} \geq 1.0 \tag{4.109}$$

Else

$$Q_g = 1.8 - \frac{0.1g}{T} \geq 1.0 \tag{4.110}$$

If the load case is classified as earthquake and the stresses in the chord result from a combination of static and spectral load cases, the spectral stress component is multiplied by a factor of 2.0. If, however, the resulting maximum stress $(f_a + f_b)$ exceeds the yield stress, the stress components, f_a, f_{ip}, f_{op} are factored such that $f_a + f_b = f_y$ and thus represent the capacity of the joint chord away from the joint (ESDEP 2017; API 2000).

4.4.2.4 Allowable Punching Shear Stress

With reference to API RP 2A-WSD, the allowable punching shear is calculated as

$$V_{pa} = \alpha Q_f Q_q \left[\frac{F_{yc}}{0.6\gamma} \right] \leq 0.4\alpha F_{yc} \tag{4.111}$$

where
F_{yc} is the yield strength of the chord member
Q_q is to account for the effects of type of loading and geometry
Q_f is a factor to account for the nominal longitudinal stress or axial load in the chord (refer to API RP 2A for the values of Q_q)

$\alpha = 1.0$ for ordinary load case
$\alpha = 1.33$ for extreme load case
$\alpha = 1.7$ for earthquake load case

The allowable punching shear is calculated separately for each component of brace loading.

4.4.2.5 Punching Shear Unity Check

With reference to API RP 2A-WSD, unity checks are calculated for each component of brace loading using the formula below (ESDEP 2017). The unity checks must not exceed 1.0 in normal conditions.

$$UC_{ax} = \left(\frac{V_P}{V_{Pa}} \right)_{ax} \tag{4.112}$$

$$UC_{ip} = \left(\frac{V_P}{V_{Pa}} \right)_{ip} \tag{4.113}$$

$$UC_{op} = \left(\frac{V_P}{V_{Pa}} \right)_{op} \tag{4.114}$$

where
   ip = in-plane bending component
   op = out-of-plane bending component
   ax = axial force component

4.4.2.6 Combined Axial and Bending Stress Unity Check

The combined axial and bending stress unity check is

$$UC_{BN} = \left(\frac{V_P}{V_{Pa}} \right)_{ip}^2 + \left(\frac{V_P}{V_{Pa}} \right)_{op}^2 \tag{4.115}$$

If $UC_{BN} > 1.0$

$$UC_{CO} = \left| \left(\frac{V_P}{V_{Pa}} \right)_{ax} \right| + \frac{2}{\pi} \arcsin \sqrt{\left[\frac{V_P}{V_{Pa}} \right]_{ip}^2 + \left[\frac{V_P}{V_{Pa}} \right]_{op}^2} \le 1.0 \tag{4.116}$$

where the arcsin term is in radians.

If the joint is specified as $C\%$ joint type 1, the axial unity check is calculated as

$$UC'_{ax} = \frac{C}{100}\left(\frac{V_i}{V_{Pa}}\right)_{ax} + \frac{100-C}{100}\left(\frac{V_r}{V_{Pa}}\right)_{ax} \qquad (4.117)$$

UC_{ip} and UC_{op} are calculated in a similar manner. The combined unity checks are calculated as before using the interpolated unity check values corresponding to each component of stress (ESDEP 2017).

4.4.2.7 Joint Strength Unity Check

The joint strength unity check is calculated by Equation 4.118.

$$UC_{jt} = \frac{f_{yb}(\gamma\tau\sin\theta)}{f_{yc}(11+1.51\beta)} \qquad (4.118)$$

4.5 DESIGN OF OTHER STRUCTURAL COMPONENTS

4.5.1 Guideposts

Guideposts are usually fixed on a radius of 12 ft with an outer diameter of 8.625 in. There are three types of guideposts (SJR 1994).

    a. Fixed. Aixed guidepost normally has a length of approximately 3 m. A fixed guidepost is used when the well slot spacing of 86 in × 86 in is required. Fixed posts are usually used together with jacket or tension leg platform tieback templates (SJR 1994).

    b. Retrievable. A retrievable guidepost normally has a length of approximately 1 m. A retrievable guidepost is used when the well slot spacing of 7 ft × 7 ft, 7 ft × 8 ft and 8 ft × 8 ft is required to suit surface tree spacing. Fixed guideposts run unto adjacent wellbays of a compact structure and will interfere with drill string tools and blowout preventers; therefore it is better to use retrievable guideposts on subsea production systems (SJR 1994).

    c. Incorporated in retrievable guide base structure.

4.5.2 Padeye and Padear

These are attachments to the primary structure provided for securing the ends of lifting slings. They are attached to the strong points of the structure, usually a node point/tubular joint. The padear is a tubular member. The analysis of the joint is that of a tubular joint. A padeye is a metal plate with a hole to accommodate a lifting pin. The padeye should be designed according to the requirement of DNV-OS-H205 and Lloyd's Register of Shipping—Code for Lifting Appliances in a Marine Environment, 2013 (Lloyd's 2013). The analyses involve sizing the padeye; equivalent stress is determined for the worst case loading by Von Mises equation.

4.5.3 Clamps

Clamps usually consist of steel shells with a lining of neoprene or similar. The clamp will normally be fabricated in two half shells and bolted together. One half shell will be fixed to a support and both half shells will be stiffened with steel plates. The following design method is based on the clamp configuration.

1. Determine the reactions (F_x, F_y, F_z) and moments (M_x, M_y, M_z) in the x, y and z planes from the analysis of the pipe that it supports.
2. Determine the clamp layout, that is, length of clamp, number of bolts, thickness of flange plates and shell.
3. From the reactions, determine the slip force and pull-off force of one half shell with respect to the other; this will dictate the bolt tension that is required and will confirm the number and size of bolts required in the design and pre-tensioning that is required.
4. Check bolt tensioning equipment around each bolt.
5. Check the hoop stress in the clamp shell.
6. Check the bearing stress in the flange plate imposed by the bolt washer.
7. Determine the flange plate thickness from an appropriate plastic hinge arrangement assuming the bolt is under its design tension.
8. Check the vertical stiffeners for in-plane bending and shear stress.
9. Analyse the clamp support for the imposed bending and axial loading.

If the clamp is designed with PTFE (polytetrafluoroethylene (more commonly known as Teflon)) bonded to the neoprene, that will allow the pipe to slip. In this case, the analysis is similar to above, but there will be no slip force in step 3. It will be necessary to establish the compressibility of the neoprene from the manufacturer to confirm that the bolt tensioning is adequate to ensure that the flange plates close together (SJR 1994).

Clamps are used in the offshore environment for retrofitting risers to jackets. They are also used on subsea structures, templates and manifolds and as a temporary installation aid for caissons, risers, pipelines, J-tubes, and so forth.

The clamp half shells are clamped onto the tubular support by pre-tensioning the stud bolts to a design pretension force.

For clamps, the only way to ensure radial pressure is maintained is to have a gap between the flange plates. As the flange plates are consequently not in compression, any additional load is transferred directly to the bolts.

The two most commonly used clamps are the anchor clamp and guide clamp. Load-bearing clamps are normally located above sea level and can either be welded directly on the jacket or be connected using a friction grip clamp. Adjustable clamps have the possibility to change the position on the riser end after installed on the jacket and are therefore more flexible.

Mainly, three analyses should be performed; ULS (ultimate limit state) stress check and FLS (fatigue limit state) check for the clamp and the bolts, in addition to a nonlinear slippage analysis. The action loads from the riser are given by the jacket operator. The environmental forces on the riser and the clamps should be calculated to find the largest load combination in the horizontal and vertical directions. For the various analyses, different load and material factors are applied to the forces. The worst load combination is applied in the analysis which is conducted in the ANSYS FE tool.

4.5.4 Plate Girder

Plate girders are typically used as long-span floor girders in buildings, as a bridge girder and as a crane girder in industrial structures. There are three types of plate girders, namely, unstiffened, transversely stiffened and longitudinally stiffened.

The main components of a plate girder are web, flanges, vertical stiffeners, longitudinal stiffeners and web/flange welds. Steel for plate girders should have sufficient notch toughness to prevent brittle fracture at the lowest anticipated ambient temperature (ESDEP 2017), (Figure 4.8).

Plate girders should be designed in accordance with the AISC Specifications for the Design, Fabrication and Erection of Structural Steel for Buildings, and Section 9 of the AWS Structural Welding Code, AWS D1.1.

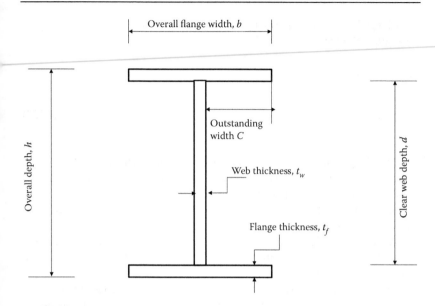

FIGURE 4.8   Plate girder geometry.

4.5.4.1 Sizing

4.5.4.1.1 Depth
Depth, h, is in the range of

$$\frac{L_0}{12} \le h \le \frac{L_o}{8} \tag{4.119}$$

where
    h = the overall girder depth
    L_o = the length between points of zero moment

For plate girder bridges,

$$h \cong \frac{L_o}{20} \tag{4.120}$$

4.5.4.1.2 Flange Breadth
Breadth, b, will usually be in the range

$$\frac{h}{5} \le b \le \frac{h}{3} \tag{4.121}$$

Flange breadth, b, is normally in multiples of 25 mm. 'Wide flats' may be used unless the flange is very wide.

4.5.4.1.3 Flange Thickness

$$\frac{c}{t_f} \leq 14\varepsilon \qquad (4.122)$$

The flange thickness, t_f, should satisfy the requirements of Table 5.3.1 of Eurocode 3 for Class 3 (semi-compact) sections. The thickness should be chosen from the standard plate thicknesses.

To ensure that beams and built-up sections will attain full plasticity and behave to the desired ductility at plastic hinge location, TM 5-1300 (1990), and AISC-WSD (1991), specify for rolled I and W shapes under compression, with width-thickness ratio for flanges not to exceed the values given in Table 4.4 (Ali 2007).

The above limiting values are from American design codes upon which most offshore designs are based. Similar codes can also be used, for example, Eurocode 3 (2005), (Ali 2007).

4.5.4.1.4 Permissible Flange Outstand, c

$$c = 12 \times t \qquad (4.123)$$

4.5.4.1.5 Web Thickness
Web thickness, t_w, will determine the exact basis for the web design, depending on whether the web is classified with regard to shear buckling as 'thick' or

TABLE 4.4   Limits of $b/2t_f$

F_Y (MPA)	$B/2T_F$
248	8.5
289	8.0
311	7.4
345	7.0
380	6.6
414	6.3
449	6.0

Source: Ali, R. M. 2007. *Performance-Based Design of Offshore Structures Subjected to Blast Loading*. London: Imperial College.

'thin' (ESDEP 2017). Thin webs will often require stiffening; this may take the form of transverse stiffeners, longitudinal stiffeners or a combination. Spacing of vertical stiffness = 0.7 d.

4.5.4.2 Design

4.5.4.2.1 Flange Plate Area

The required flange plate area, A_f, may readily be obtained as follows:

$$A_f = \frac{M}{[(h - t_f)f_y/\gamma MO]} \simeq \frac{M}{(hf_y/\gamma MO)} \tag{4.124}$$

(An iteration or two may be required depending on an assumed value of t_f and its corresponding f_y value from Table 3.1, Eurocode 3). The flange b/t_f ratio need only comply with the Eurocode 3 (Table 5.3.1) requirements for a Class 3 (semi-compact) flange (ESDEP 2017).

4.5.4.2.2 Cross-Sectional Moment of Resistance

The cross-sectional moment of resistance may then be checked using Equation 4.125.

$$M_{f.Rd} = \frac{bt_f(h - t_f)f_y}{\gamma MO} \tag{4.125}$$

where $M_{f.Rd}$ = cross-sectional moment of resistance.
All other parameters are as defined above.

4.5.4.2.3 Web Plate

Web plate can be sized by assuming uniform shear stress τ_y over its whole area. The web-to-flange fillet welds must be designed to transmit the longitudinal shear at the flange–web interface.

4.5.4.2.4 Shear Stress

Average shear stress is calculated as

$$\frac{V}{t_w \times d_w} \tag{4.126}$$

4.5.4.2.5 Buckling

Once the d/t_w value for an unstiffened web exceeds a limiting value (69ε in Eurocode 3), the web will buckle in shear before it reaches its full shear capacity $A_w\tau_y$.

Diagonal buckles resulting from the diagonal compression associated with the web shear will form. Their appearance may be delayed through the use of vertical stiffeners. The load at which shear buckling is initiated is a function of both d/t_w and panel aspect ratio a/d.

Provided that outstand proportions c/t_f are suitably restricted, local buckling will have no effect on the girder's load-carrying resistance.

Webs for which $d/t_w \leq 124\varepsilon$ and which are not subject to any axial load will permit the full elastic moment resistances of the girder to be attained. If this limit of d/t_w (or a lower one if axial compression in the girder as a whole is also present) is exceeded, then moment resistance must be reduced accordingly. If it is desired to reach the girder's full plastic moment resistance, a stricter limit will be appropriate (ESDEP 2017).

If particularly slender webs are used, the compression flange may not receive enough support to prevent it from buckling vertically, rather like an isolated strut buckling about its minor axis. This possibility may be eliminated by placing a suitable limit on d/t_w. Transverse stiffeners also assist in resisting this form of buckling.

Vertical loads may cause buckling of the web in the region directly under the load as for a vertical strut. The level of loading that may safely be carried before this happens will depend upon the exact way in which the load is transmitted to the web, the web proportions, and the level of overall bending present (ESDEP 2017).

4.5.5 Well Conductor

Conductor is a tubular pipe extending upward from or beneath the seafloor containing pipes that extend into the petroleum reservoir (ISO 19900:2013).

A conductor is generally vertical and is continuous from below the seafloor to the wellbays in the topsides and can be laterally supported in both the support structure and topsides structure. The vertical support is in the seabed.

In a few cases, conductors are rigidly attached to the topsides or to the support structure above sea level. In these cases, the conductor's axial stiffness can affect the load distribution within the overall structure.

Conductors are long, hollow, straight or curved tubes that embed into the seabed through which drilling is performed. To support such a long length of the tube, conductor framings are provided (Nallayarasu 2013).

4.5.6 Pipeline and Umbilical Interfaces

The template may have to make provision for connection and possible disconnection of pipelines, flowlines and control lines at a series of porches on the template structure.

The porch is the fixed end of a connection specially designed to fit in a connection system. It includes the pipeline end (hub) which is fixed to the porch. The layout and configuration of the porches will depend on the layout and spacing of the wellheads, manifold pipework and control system on the proposed method of connection (SJR 1994). Typical features of porches include

• Ramp to assist pull-in of the outboard connector skid assembly by either diverless or diver-assisted techniques.

• Prefabricated porch structure complete with guidepost arrangements, inboard connector, locking mechanisms, and so forth, for pull-in and connection by diverless techniques.

• Inboard connector, work platform, holding points, lifting frames, and so forth, for pull-in and connection by divers.

• Means of preventing drilling cuttings and other debris from accumulating on the seabed in front of the porches, or on the porches themselves. This generally takes the form of a solid bulkhead between the porch and the adjacent well slots.

The porch structure and the connector should be capable of transmitting the following loads to the main template structure:

• Reactions due to pull-in operations
• Reaction due to connector clamping operations
• Loads due to lengths of flow and control lines unsupported by the seabed, including perhaps an allowance for scouring
• Loads due to thermal expansion of pipeline and flowline spool pieces and jumpers

The above loads may be fairly accurately defined once details of the lines and the connection method are known.

Breakaway loads due to anchor dragging of flow and control lines cannot be easily defined. Some clients incorporate flow and control line load-limiting devices adjacent to the template, which cause the line to break at a specified tensile load. In such a case, the template must be designed to withstand the specified tensile load (SJR 1994).

4.6 FIRE DESIGN OF STRUCTURAL CONNECTIONS

The design of connections for fire safety involves calculation of the heat transfer to them and a determination of the response of the structural elements of the connections. PrEN 1993-1-2 (2003) gives two methods for the design of steel connections. The first method involves applying fire protection to the member and its connections. The level of protection is based on that applied to the connected members taking into account the different level of utilisation that may exist in the connection compared to the connected members (Moore and Wald 2003). For the second method, we use an application of the component approach in prEN 1993-1-8, together with a method for calculation of the behaviour of welds and bolts at an elevated temperature. By using this methodology, the connection moment, shear and axial capacity can be calculated at an elevated temperature (Moore and Wald 2003).

For the fire design of a structural connection, the following needs to be considered

a. Bolt resistance at high temperature
b. Weld resistance at high temperature
c. Temperature distribution with time within a joint
d. Fire resistance of joints

The methodology for the fire design of structural connections is available from prEN 1993-1-2, prEN 1993-1-8 (Moore and Wald 2003).

Impact Analysis

5

The platform should survive the initial impact from dropped objects and meet the post-impact criteria as defined for vessel collision. Providing a singularly strong element to resist impact loads does not guarantee global impact resistance; the entire framework must be arranged to confine and control damage under the full range of plausible impact conditions.

To lower the probability of offshore platform collision by passing vessels, the following schemes are suggested.

a. Fendering. Fendering should be used to protect the exposed parts of the installation, such as risers and boat landings, against minor collisions.

The most commonly used fenders on steel installations are fixed bumpers covered with car tyres or similar.

The effect of fendering as a collision prevention scheme has been excluded from some studies; fendering is unlikely to protect large possible damage risk areas.

If fendering is provided to protect parts of the structure against impact of supply vessels, then the fendering should be capable of absorbing the impact energy of the vessel. It should be designed not to become detached from the main structure or the vessel striking it in the event of overload. Damage to the chord should be avoided by the use of fendering, since a reduction in the strength of a leg of the jacket cannot be designed against.

b. Radar. Install automatic warning radar on platforms or guard ships, together with the means to alert approaching ships, such as the use of flares, foghorns and high-power lights. Radar transponders may reduce the need for the ship to pass close to the platform by marking and identifying a fixed installation. Radar and radar echo-enhancement devices can be used as collision-avoidance aids. Radar corner reflectors can be used to mark small structures or vessels that may not be clearly visible on a ship's radar.

c. Sound and light signals on platforms.

d. Guard ships/support vessels.

e. Standby vessels can be equipped with measures to alert vessels, such as flares, foghorns and high-power lights. These should be used in conjunction with existing measures such as contact via VHF and MF radio. Helicopters, if available, could be sent to alert ships that are threatening collision. Calling vessels on VHP/radio and other communication systems can also be employed.

f. Increased radius of safety zone. A safety zone of 500 m is normally used. However, there is a need for a strong surveillance of the safety zone and judicial prosecution of violations need to be enforced.

g. Establishment of ship corridors. Flashing and auditory navigation buoys should be installed some miles in advance of the platform. This will alert both the ship and the platform some time before a collision. Employ traffic separation schemes to minimize congestion-related collisions.

h. Use of RACON on platforms.

i. Improved distribution of information about the platform.

j. Installation of automatic collision-avoidance systems on board.

k. Inclusion of the ship/platform collision-avoidance knowledge and skill training into mariners training, education and examination.

l. Developmetn of ship/platform collision treatment systems and installation on platforms.

m. Set up of ship routing systems if a platform is located in or quite close to a high traffic density shipping lane.

n. Installation of collision risk auto warning system on platforms.

5.1 IMPACT ENERGY

The impact energy calculation is based on the requirements of API RP 2A. The kinetic energy of a vessel can be calculated using Equation 5.1.

$$E = 0.5amv^2 \tag{5.1}$$

where

E = kinetic energy of the vessel (impact energy)

a = added mass factor ($a = 1.4$ for broadside collision, $a = 1.1$ for bow/stern collision)

m = vessel mass

v = velocity of vessel at impact

$m = 1000$ tonnes for platforms in mild environment and close to their base of supply

$v = 0.5$ m/s

The API RP 2A code includes the effect of impact caused by a vessel berthing against a platform in the assessment of dynamic loads. Otherwise, it does not consider further any accidental collisions and damage.

With reference to BS 6235, the minimum impact to be considered is of a vessel of 2500 tonnes at 0.5 m/s. In the absence of a suitable computational method, it should be assumed that all of the energy is absorbed by the structure and none by the ship.

5.2 DAMAGE ASSESSMENT

The force required to locally damage tubular members can be calculated using Equation 5.2 below.

$$P_d = 15 M_p \left(\frac{D}{t} \right)^{1/2} \left(\frac{X}{R} \right)^{1/2} \tag{5.2}$$

where

P_d = the denting force
M_p = the plastic moment capacity of the tube
$M_p = F_y t^2 / 4$ with F_y being the yield strength
D = the diameter of the tube
R = the radius of the tube
t = the wall thickness
X = the dent depth

5.3 BENDING AND DEFLECTION CHECK

The design of members subjected to lateral impact loading includes bending and deflection checks. The impact analysis below is applicable to propped cantilever, simply supported, fixed ended members, tubular members, grating members, and so on (SJR 1994).

Assumptions

1. Assumes that the cross-section of the member is and remains undamaged at the time of impact.
2. Assumes the cross-section of the member is compact, that is, is capable of developing full plastic moment of resistance.
3. Assumes that all the energy of impact is absorbed by the development of plastic hinges in an impacted member. For long members, a significant amount of impact energy might be absorbed by elastic bending.
4. Is a conservative approximation.
5. Is suitable for the design of protection covers and members.
6. Compares the energy absorption capacity of a member with the energy of an impacting load.

5.3.1 Bending

$$M_P = F_Y Z_P \tag{5.3}$$

$$M_E = F_Y Z_E \tag{5.4}$$

$$M_E = \frac{M_P Z_E}{Z_P} \tag{5.5}$$

$$\frac{0.75}{M_P} = \frac{L_P}{M_P - M_E} \tag{5.6}$$

$$L_E = \frac{0.75 L (Z_P - Z_E)}{Z_P} \tag{5.7}$$

Maximum acceptable strain = 22% (refer to BS 4360).
Limit strain to 9%:

$$x = 9\% = D\theta 100/L_P$$

$$\theta = \frac{0.09 L_P}{D} \tag{5.8}$$

$$E_R = 3M_P \theta \tag{5.9}$$

$$\therefore E_R = \frac{0.27L_P T_Y Z_P}{D} \tag{5.10}$$

$$\therefore E_R = 0.2L(Z_P - Z_E)F_Y D \tag{5.11}$$

$$E_I = M \times g \times V \tag{5.12}$$

If $E_R > E_I$, then the member passes the bending check for impact load.

5.3.2 Deflection

$$E_I = 3M_P \theta$$

$$\theta = \frac{E_I}{3M_P} RAD$$

$$\theta = \frac{E_I 360}{6\pi M_P} \text{degrees}$$

$$\Delta = \frac{L}{2}\sin\theta \tag{5.13}$$

where
  M = mass of impacting object (kg)
  v = velocity of impacting object (m/s)
  E = energy (kJ)
  E_I = impact energy (kJ)
  E_R = resistance energy
  M_P = plastic moment of resistance (kN m)
  M_E = elastic moment of resistance (kN m)
  L = span of member (m)
  θ = joint rotation radius
  x = strain
  B = width of member (m)
  D = depth or diameter of member (m)
  F_y = yield stress (kN/m²)
  Z_P = plastic modulus (cm³)
  Z_E = elastic modulus (cm³)
  g = acceleration due to gravity (m/s²)
  Δ = deflection

5.3.3 Example of an Impact Analysis of a Tubular Member

$197.3 OD \times 36.6 \, kg/m$, $Z_E = 252 \, \text{cm}^3$, $Z_p = 338 \, \text{cm}^3$, $L = 6 \, \text{m}$

Bending:

$$F_y = 275,000 \, \text{kN/m}^2, \quad M = 8 \, \text{kg}, \quad v = 0.25 \, \text{m/s}$$

$$E_1 = 8 \times 9.81 \times 0.25 \cong 20 \, \text{kJ}$$

$$x = 9\% \text{ max strain}$$

$$E_R = 0.2L(Z_p - Z_E)F_y/D$$

$$E_R = 0.2 \times 6(338 - 252)275000/193.7 \times 10^3$$

$$E_R = 146.5 \, \text{kJ}$$

$E_R > E_1$, therefore the tubular member passes the bending check for an impact load.

5.4 PUSHOVER ANALYSIS

Pushover analysis of a platform demonstrates that the platform has adequate strength and stability to withstand the loading criteria. Pushover analysis is well suited for static loading, ductility analysis or dynamic loading. An example of such loading includes waves acting on a stiff structure with natural periods under three seconds or ice loading. The pushover analysis can be used for ductility level earthquake analysis by demonstrating that the platforms strength exceeds the maximum loading for the extreme earthquake events (NORSOK 2008). Post-impact strength analysis is carried out to check the structure against 1-year environmental loads with all dead loads. This is to ensure the structure is able to function as normal for some period of time so that any repair needs to be carried out due to the impact pushover analysis is carried out to check the global integrity in terms of collapse behaviour.

The finite element model program called ABAQUS or SACS can be used for the pushover analysis of the offshore platform. The analysis should be performed using the full nonlinear structural model. ABAQUS uses full Newton-Raphson iteration to determine the nonlinear response of the platform. Geometric nonlinearity will be included and the load-displacement curves should be obtained using the automatic increment options, based on the modified risks method (NORSOK 2008).

In conducting the analysis, first apply the dead loads and the buoyancy loads and then incrementally increase until it reaches the extreme load specified. As the load is increased, structural members such as joints, or piles, should be checked for inelastic behaviour to ensure proper modelling.

Lifting Analysis

6

6.1 LIFTING ARRANGEMENT

Offshore platforms are normally lifted by cranes using a three- or four-sling, single-hook arrangement. In certain situations, a two-hook/tandem lift, a lifting frame or lifting bars may be used.

A three-sling lifting arrangement is statically determinate and is moderately unaffected by small deviations from an ideal sling. Computer analysis of a three-sling lift arrangement does not consider the effect of skew loading/short slings. Therefore, only four-sling lifts will be considered.

A four-sling arrangement is not statically determinate. The stress/strain relationship between the slings and structure during lift is not linear. Deviations from the ideal sling length may have a marked effect on the whole system. A sling tolerance factor should be incorporated into the lift calculation/analysis.

A short sling causes a structure to span between the corner common to the short sling and the diagonally opposite corner.

The minimum sling angle is 60° to the horizontal. The greater the sling angle, the less the loading on the lifted structure.

Lifting points should be provided with lifting lugs. The lifting lugs should be designed to provide easy attachment and removal of slings. The lugs are usually padeyes or padears. The sling length should be selected such that before lifting, the hook is vertically over the centre of gravity of the jacket being lifted to avoid possible damage to the jacket and/or barge during the lifting process (GPB 2010; Lloyd's 2013; API 2000).

6.2 WEIGHT CALCULATION

Allow 3% of W for weight measurement.
Allow 10% of W for weight contingency.

$$\text{Load factor } (LF) = 0.13 + \text{Rigging load factor} \qquad (6.1)$$

where, W is the weight of the structure.

6.2.1 Lifting Forces

6.2.1.1 Lifting Force Calculation Based on API RP 2A

For lifts to be made at offshore locations, padeyes and other internal members (and both end connections) framing into the joint where the padeye is attached and transmitting lifting forces within the structure should be designed for a minimum load factor of 2.0 applied to the calculated static loads (API 2000). All other structural members transmitting lifting forces should be designed using a minimum load factor of 1.35.

For other marine situations (i.e. loadout at sheltered locations), the selection of load factors should meet the expected local conditions but should not be less than a minimum of 1.5 and 1.15 for the two conditions (refer to Section 2.4.2 of API RP 2A).

For typical fabrication yard operations in which both the lifting derrick and the structure or components to be lifted are land-based, dynamic load factors are not required. For special procedures in which unusual dynamic loads are possible, appropriate load factors may be considered (API 2000).

Fabrication tolerances and sling-length tolerances both contribute to the distribution of forces and stresses in the lift systems which are different from that normally used for conventional design purposes. The load factors recommended are intended to apply to situations where fabrication tolerances do not exceed the requirements, and where the variation in the length of slings does not exceed plus or minus one-fourth of 1% of nominal sling length, or 1.5 in.

The total variation from the longest to the shortest sling should not be greater than one-half of 1% of the sling length, or 3 in. If either fabrication tolerance or sling-length tolerance exceeds these limits, a detailed analysis, taking into account these tolerances, should be performed to determine the redistribution of forces on both slings and structural members. This same type of analysis should also be performed in any instances where it is anticipated that unusual deflections of particularly stiff structural systems may also affect load distribution (API 2000).

Consequently, at the padeye location,

$$LF = 1.13 \times 2 = 2.26 \tag{6.2}$$

and, remote from padeye,

$$LF = 1.13 \times 1.35 = 1.526 \cong 1.53 \tag{6.3}$$

6.2.1.2 Lifting Force Calculation Based on Noble Denton's Requirements

If the lifting operation will not take place under adverse conditions, the dynamic amplification factors (DAFs) given in Table 6.1 may be considered (GPB 2010).

6.2.1.2.1 Skew Load Factor (SKL)

With reference to Noble Denton's guidelines for marine lifting operations, the skew load is a load-distribution factor based on sling-length manufacturing tolerances, rigging arrangement and geometry, fabrication tolerances for lift points and sling elongation and should be considered for any rigging arrangement and structure that is not 100% determinate (GPB 2010).

A significantly higher SKL factor may be required for new slings used together with existing slings, as one sling may exhibit more elongation than the others.

TABLE 6.1 In-air dynamic amplification factors

| GROSS WEIGHT, W (TONNES) | DYNAMIC AMPLIFICATION FACTOR (DAF) | | | |
| | | | ONSHORE | |
	OFFSHORE	FLOATING INSHORE	MOVING	STATIC
$W \leq 100$	1.30	1.15		1.00
$100 < W \leq 500$	1.25	1.10		1.00
$500 < W \leq 1000$	1.20	1.10		1.00
$1000 < W \leq 2500$	1.15	1.05		1.00
$2500 < W \leq 10,000$	1.10	1.05		1.00

Source: GPB. 2010. *Guidelines for Marine Lifting Operation – 0027/ND*. Noble Denton Group Limited.

For indeterminate four-sling lifts using matched pairs of slings, an SKL of 1.25 shall be applied to each diagonally opposite pair of lift points in turn.

For determinate lifts, the SKL may be taken to be 1.0, provided it can be demonstrated that sling-length errors do not significantly affect the load attitude or lift system geometry.

For a lift system incorporating spreader bars using matched pairs of slings, an SKL of 1.10 is applicable.

For a lift system incorporating a single spreader bar using matched pairs of slings, an SKL of 1.05 is applicable.

For multi-hook lifts where the hook elevation can be shown to be individually controlled, a lower SKL than stated above may be applicable, subject to evaluation of sling-length tolerances, rigging arrangement and crane operating procedures.

For a single-hook lift where four slings of unequal length are used (i.e. not matched pairs), the SKL shall be calculated by the designer (considering sling-length tolerances and measured lengths) and applied to the structure and lift system design accordingly. Where the calculated SKL is less 1.25, an SKL of 1.25 shall be applied (GPB 2010).

6.2.1.3 Lifting Force Calculation Based on Lloyd's Register Requirements

The structure should be designed for the balanced lift case and unbalanced lift case.

a. Balanced lift. The lifting sling origin coincides, in the plan view, with the centre of gravity of the structure, with the padeye loads taken to be direct function of their distance from the centre of gravity. The load factor of 2.0 is to be applied to all gravity loads used for structural analysis purposes (Lloyd's 2013).

$$\text{Load factor } (LF) = 1.13 \times 2 = 2.26 \tag{6.4}$$

b. Unbalanced lift. For unbalanced lift, a load factor of 1.33 is to be applied to all gravity loads for structures having a lift weight up to 500 tonnes. For heavier lift, consider reducing this factor. However, this factor should never be less than 1.10.

75% of the lifting weight should be carried by two diagonally opposite padeyes.

25% of the lifting weights should be carried by the other two padeyes.

$$\text{Load factor } (LF) = 1.13 \times 1.33 = 1.50 \tag{6.5}$$

Transportation Analysis 7

7.1 DATA REQUIRED FOR ANALYSIS

a. Details of seafastening locations
b. Details of transportation vessel
c. Details of the structure and equipment, for example, layout, self-weight, equipment weight, and so forth
d. Wind data
e. Details of dead weight allowance for contractors' equipment that will be transported with structure, for example, rigging
f. Vessel/environmental design criteria
g. An environmental report showing the worst sea-state conditions during that time of the year throughout the course of the intended route should be available for design

7.2 TRANSPORTATION ANALYSIS OF MOTION FORCE

The following parameters are needed for the motion analysis:

- Significant wave height representative of the tow route
- A range of peak wave periods
- Wind speed
- Vessel heading relative to the waves

Range of peak wave period

$$\sqrt{13H_s} \leq T_p \leq \sqrt{30H_s} \qquad (7.1)$$

where H_s is expressed in metres and T_p in seconds.

The design wave height, H_s, can be based on a 10-year return, adjusted for the period of exposure. The range of peak wave period can be used to account for different wave steepness and can be calculated using the expression above. If the peak roll period of the barge falls outside the T_p range for the design wave, smaller waves with periods smaller to that of the barge roll period are also analysed.

The transportation analysis is to derive the inertia force that may be imposed on a structure when transported by sea and the use of the inertia force in the analysis of the structure. The angular rotation (θ) is termed as roll or pitch and the vertical movement is termed as heave (SJR 1994), (Figure 7.1).

During transportation, a ship and its cargo are subject to accelerations and retardations due to wind and the sea state. The inertia of an item is calculated as

$$F = ma \tag{7.2}$$

where

F = inertia force
m = mass of the item
a = acceleration

The inertia force (F_t) of an element can be set into components to simplify computer input and final analysis, that is

$$F_t = \sqrt{\left(F_x^2 \quad \text{or} \quad F_y^2\right) + F_z^2} \tag{7.3}$$

where x, y and z represent the structure/cargo global axis.

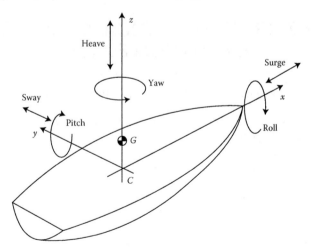

FIGURE 7.1 Degrees of freedom of a vessel.

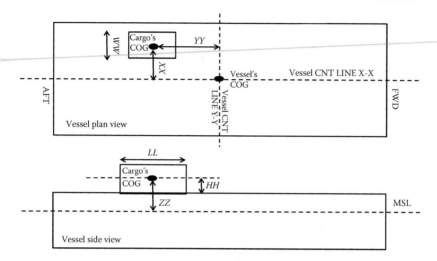

FIGURE 7.2   Plan view of a vessel.

Assuming sinusoidal wave actions or simple harmonic motions and based on standard text *Handbook of Offshore Engineering* by Subrata Chakrabarti (2005), the following calculation is made (see Figure 7.2).

Design horizontal force, roll and heave effect, F_{HR},

$$F_{HR} = Wt \left[\frac{4\pi^2 R\theta ZZ}{T^2 g} + (1 + H_g)\sin R\theta \right] \tag{7.4}$$

Design vertical force, roll and heave, F_{VR},

$$F_{VR} = Wt \left[\frac{-4\pi^2 R\theta XX}{T^2 g} + (1 + H_g)\cos R\theta \right] \tag{7.5}$$

Overturning force, roll and heave effect, F_{OTR},

$$F_{OTR} = \frac{F_{HR} \times HH}{WW} \tag{7.6}$$

Resistance force, roll and heave effect, F_{RR},

$$F_{RR} = \frac{F_{VR}}{2} \tag{7.7}$$

Design horizontal force, pitch and heave effect, F_{HP},

$$F_{HP} = 1.2Wt\left[\frac{4\pi^2 ZZP\theta}{T^2 g} + (1 + H_g)\sin P\theta\right] \tag{7.8}$$

Design vertical force, pitch and heave effect, F_{VP},

$$F_{VP} = Wt\left[\frac{-4\pi^2 YYP\theta}{T^2 g} + (1 + H_g)\cos P\theta\right] \tag{7.9}$$

Overturning force, roll and heave effect, F_{OTP},

$$F_{OTP} = \frac{F_{HP} \times HH}{LL} \tag{7.10}$$

Resistance force, pitch and heave effect, F_{RP},

$$F_{RP} = \frac{F_{VP}}{2} \tag{7.11}$$

Rotational moment of inertia, M_R,

$$M_R = I_{yy}\left[\pm\frac{4\pi^2\theta}{T^2 g}\right] \tag{7.12}$$

Uplift

Roll and heave: If $OTF_r > RF_r$, Uplift

Pitch and heave: If $RF_p > OTF_p$, Uplift

where
$R\theta$ = roll angle (radians)
$P\theta$ = pitch angle (radians)
H_g = heave (g) $(H_g = 0.2)$
T = roll, pitch or heave period (sec)
DH = barge draft (m)
D = barge depth (m)
FD = freeboard (m)
ZZ = height above the centre of rotation (assumed to be at waterline)

YY = horizontal distance from the centre line of the barge
g = gravitational acceleration (m/s^2)
θ = roll or pitch angle (radians)
I_{yy} = moment of inertia of the cargo about its longitudinal axis
FHR = inertia force parallel to the vessels deck
FVR = inertia force normal to the vessel deck
O = an element of the structure
W = self-weight of element O

In the absence of details of the transportation vessel, the following parameters in Table 7.1 may be assumed for a first approximation.

7.3 SEAFASTENING

Before transporting the platform, seafastening analysis is done and the platform parts (decks, jacket and appurtenances) are fastened to the barge. Seafastening includes any grillage, dunnage, cribbing or other supporting structure, roll, pitch and uplift stops, and the connections to the barge or vessel. Seafastening involves fitting and welding sufficient structure between the structure and the barge to prevent the jacket from shifting during transit to the offshore site.

The design of the grillage and seafastening should be carried in accordance with the requirement of the AISC and the API RP 2A. For the design of the seafastening members, allowable stress may be increased by multiplying it with a factor; API RP 2A states that for temporary and transient load conditions, one-third increase is allowable (SJR 1994). This is to reflect the transient and extreme nature of the transport load. The third increase in the allowable stresses does not, however, apply to the local strength of the deck of

TABLE 7.1 Assumed parameters for sea transportation analysis

	SINGLE AMPLITUDE (10-SECOND PERIOD)		
	ROLL (°)	PITCH (°)	HEAVE (G)
Ships	30	15	0.2
Cargo (<250 ft LOA) Barges (<75 ft beam)	25	15	0.2
Barges (larger)	20	12.5	0.2
Barges (inland tows)	5 (1)	5	0.1

LOA = Length Overall of the vessel.

the carrier vessel. Refer to Section 3.5 of DnV-OS-H202 (2016) for information on grillage seafastening.

Dog plates (clips) are seafastening items which are used to restrain the cargo from tipping during transport. All dogs shall be in good operating condition and seals shall be functioning correctly. Stress check is as per the requirement of AISC specification for structural steel buildings—allowable stress design (ASD).

Lashing is used to secure cargo for transportation, with the goal of minimizing shifting. Items used for lashing include ropes, cables, wires, chains, strapping, shackles, turnbuckles and nets. These items are anchored to the container and tensioned against the cargo. Another form of lashing used four devices attached to the top of each corner of a container. For lashing seafastening refer to Section 6.2.4 of DNV-OS-H202.

The design load in any chain, wire or webbing strap used for seafastening should not exceed the certified (lifting) working load limit (WLL) or safe working load (SWL) of the seafastening. When the WLL or SWL is not known, the design load shall be taken as no more than the certified breaking load (BL) divided by 2.25.

Wire/cargo strap SWL is calculated as

$$SWL = \frac{\text{Minimum breaking load } (BL)}{\text{Safety factor } (SF)} = \frac{BL}{2.25} \tag{7.13}$$

where BL = certified minimum breaking load of wire rope, chain or shackles.

There are spreadsheets available for seafastening designs of grillage, dog plate and lashing.

Fatigue Analysis

8

Fatigue is the weakening of a material caused by the repeated application and removal of stress. For example, if you bend a piece of metal back and forth repeatedly in the same spot, fatigue will result at the bend location and will weaken the metal until it eventually breaks.

Technically, platforms experience fatigue as a result of periodic increases (application of stress) and decreases (removal of stress) in operating pressures. Because fatigue can cause a failure to occur at stress levels well below those that a material can withstand in a single, nonrepetitive loading, materials that must resist repeated stress cycles must be specially designed for this service. Durability is the ability of the material to resist fatigue.

Fatigue analysis is done to ascertain the offshore platform structural response to continual wave loading. Fatigue analysis is carried out using the deterministic method or spectral method. Use the $S-N$ curves to determine fatigue cycles. Also calculate the fatigue stress or fatigue damage.

A detailed fatigue analysis should be performed for all structures as recommended in Section 5.2 of API RP 2A.

The fatigue analysis is performed with input from a wave scatter diagram and from the natural dynamic response of the platform and the stiffness of the pile caps at the mud line by applying the Palmgren-Miner formula.

Miner's rule is probably the simplest cumulative damage model. It states that if there are k different stress levels and the average number of cycles to failure at the ith stress, S_i, is N_i, then the damage fraction, C, is

$$C = \sum_{i=1}^{k} \frac{n_i}{N_i} \tag{8.1}$$

where
n_i = the number of cycles accumulated at stress S_i
C = is the fraction of life consumed by exposure to the cycles at the different stress levels. In general, when the damage fraction reaches 1, failure occurs

A detailed fatigue analysis should be performed to assess cumulative fatigue damage.

8.1 FATIGUE LIFE

8.1.1 Cumulative Fatigue Damage

The wave climate should be derived as the aggregate of all sea states to be expected over the long term. A spectral analysis technique should be used to determine the stress response for each sea state. Dynamic effects should be considered for sea states having significant energy near a platform's natural period. Local stresses that occur within tubular connections should be considered in terms of hot-spot stresses located immediately adjacent to the joint intersection using suitable stress concentration factors. The stress responses should be combined into the long-term stress distribution, which should then be used to calculate the cumulative fatigue damage ratio, D, giving adequate consideration to both global and local stress effects (Nallayarasu 2013; API 2000).

$$D = \sum \left(\frac{n}{N} \right) \tag{8.2}$$

where

n = number of cycles applied at a given stress range
N = number of cycles for which the given stress range would be allowed by the appropriate S–N curve

When fatigue damage can occur due to other cyclic loadings, such as transportation, the following Equation should be satisfied

$$\sum_j SF_j D_j < 1.0 \tag{8.3}$$

where

Dj = the fatigue damage ratio for each type of loading
Sfj = the associated safety factor. Safety factors can be obtained from API RP 2A

8.1.2 Stress Concentration Factors

There is a high concentration of local stresses at the welds of tubular joints. These welds are the most fatigue-sensitive areas in an offshore platform. Fatigue life should be estimated by calculating the 'hot spot stress range'. The hot spot stress range ($HSSR$) is then used as an input to the suitable S–N curve. For each tubular joint configuration and each type of brace loading, the stress concentration factor (SCF) is defined as

$$SCF = HSSR/\text{Nominal brace stress range} \qquad (8.4)$$

The nominal brace stress range should be based on the section properties of the brace end under consideration (Nallayarasu 2013; API 2000). Use the Efthymiou equation to calculate the SCF of unstiffened welded tubular joints. Depending on the joint configuration, brace under consideration, and loading pattern, classify the tubular joint as T/Y, X, K or KT.

Apply the Lloyd reduction factors to the $SCFs$ for the equivalent unstiffened joint to evaluate the $SCFs$ for internally ring-stiffened joints.

For cast joints, SCF is calculated as the maximum principal stress at any point on the surface of the casting divided by the nominal brace stress outside the casting.

8.1.3 Curves

For a particular stress range 'S', there exists a theoretical number of cycles 'N' at which fatigue failure may occur. The relationship between this number of allowable cycles and the stress range is usually expressed as an S–N curve.

For welded tubular and cast joints, the basic design S–N curve is of the form

$$\log_{10}(N) = \log_{10}(k_1) - m\log_{10}(S) \qquad (8.5)$$

where

N = the predicted number of cycles to failure under stress range S
k_1 = a constant
m = the inverse slope of the S–N curve

Table 5.5.1-1 of API RP 2A presents the basic welded tubular joints and cast joints S–N curves. The material-thickness effect for welded joints and castings should be estimated and applied where applicable (Nallayarasu 2013; API 2000).

S–N curves for various steel sizes are obtainable from API RP 2A-WSD or ISO 19902.

Dynamic Analysis

9

A dynamic analysis of an offshore platform is indicated when the design sea state contains significant wave energy at frequencies near the platform's natural frequencies. In dynamic analysis, the natural periods are further used in the seismic analysis and spectral fatigue analysis.

Dynamic analysis covers the behaviour of structures subjected to dynamic (actions having high acceleration) loading. Dynamic loads include people, wind, waves, traffic, earthquakes and blasts. Any structure can be subjected to dynamic loading. Dynamic analysis can be used to find dynamic displacements, time history and modal analysis.

A dynamic analysis is also related to the inertia forces developed by a structure when it is excited by means of dynamic loads applied suddenly (e.g. wind blasts, explosion, and earthquake).

A static load is one which varies very slowly. A dynamic load is one which changes with time fairly quickly in comparison to the structure's natural frequency. If it changes slowly, the structure's response may be determined with static analysis, but if it varies quickly (relative to the structure's ability to respond), the response must be determined with a dynamic analysis.

Dynamic analysis for simple structures can be carried out manually, but for complex structures, finite element analysis can be used to calculate the mode shapes and frequencies (Wikipedia 2017).

9.1 EQUATION OF MOTION

The horizontal displacement of the platform is represented by X measured with respect to the neutral position of the centre of gravity of the deck (Harleman 2010; NORSOK 2008). The equation of motion for a multi-degrees-of-freedom system can be expressed in matrix form:

$$Mx'' + CX' + KX = P(t) \tag{9.1}$$

$$m\frac{d^2X}{dt^2} + C\frac{dX}{dt'} + KX = P(t') = P_m \sin \sigma t' \qquad (9.2)$$

The horizontal displacement of the platform is calculated using an equation proposed by Housner and Hudson (Harleman 2010) as

$$X(t') = \frac{P_m}{k\sqrt{[1-(\sigma/\sigma_n)^2]^2 + [2(C\sigma/C_c\sigma_n)]^2}} \sin(\sigma t' - \Phi) \qquad (9.3)$$

where

$P(t')$ = the harmonic exciting force
m = effective mass of the system
C = damping coefficient of structure
K = stiffness matrix or K = the spring constant of the system
P_m = amplitude of harmonic exciting force
σ = frequency of harmonic exciting force = $2\pi/T$
t' = time in the equivalent system
T = period of the wave equation
$X(t')$ = horizontal displacement due to $P(t')$
σ_n = undamped natural frequency of spring-mass system

$$\sigma_n = \sqrt{\frac{K}{m}} \qquad (9.4)$$

C_c = critical damping coefficient

$$C_c = 2\sqrt{mK} \qquad (9.5)$$

Φ = phase angle

$$\Phi = \tan^{-1}\left[\frac{2(C/C_c)\cdot(\sigma/\sigma_n)}{1-(\sigma/\sigma_n)^2}\right] \qquad (9.6)$$

$$\frac{C}{C_c} = \frac{1}{2\pi}\left(\ln\frac{X_f}{X_2}\right) \qquad (9.7)$$

$P(t)$ is the time-dependent force vector; in the most general case, it may depend on the displacements of the structure also (i.e. relative motion of the structure with respect to the wave velocity in Morison equation) (Harleman 2010).

The total platform displacement as a function of time is calculated using Equation 9.8.

$$X_{tot}(t) = \sum_{m=0}^{m} X_m(t) \tag{9.8}$$

The usual procedures of structural analysis may then be used to relate displacements (strains) to stresses.

The static deflection of the platform is obtained from cantilever beam theory (Harleman 2010; Visser 1995). The equation for the maximum deflection of the deck (X_{max}) for a force, F, applied at an elevation, $\bar{s}$, is

$$[X_{max}]_{act} = \frac{F\bar{s}^2}{NEI}\left(\frac{L}{4} - \frac{\bar{s}}{6}\right) \tag{9.9}$$

where

F = force applied at the elevation of the deck
E = elastic modulus of actual platform leg
I = moment of inertia of actual platform leg
N = number of platform legs
L = vertical length of legs
$\bar{s}$ = elevation

The natural frequency, σ_n, is calculated using Equation 9.10.

$$\sigma_n = \sqrt{\frac{K}{1/g(W + (13/35)NwL)}} \tag{9.10}$$

where, W is the weight of deck and w is the weight per foot of leg.
Effective mass of the platform, m, is calculated as

$$m = \frac{1}{g}\left(W + \frac{13}{35}NwL\right) \tag{9.11}$$

The critical damping coefficient, C_c, is calculated as

$$C_c = 2\sqrt{mK} \tag{9.12}$$

9.2 DYNAMIC AMPLIFICATION FACTORS

Dynamic amplification factor (DAF) is used to measure the dynamic effects of a structure subjected to dynamic loads. The DAF may be defined as the ratio of the (maximum) dynamic response to the static response of the structure by considering the loads as dynamic and quasi-static, respectively.

For rigid structures having a fundamental vibration period well below the range of wave periods (typically less than 3 seconds), the dynamic behaviour is simply accounted for by multiplying the time-dependent loads by a dynamic amplification factor (Harleman 2010; NORSOK 2008; Visser 1995).

$$DAF = \frac{1}{\sqrt{\left(1-\left(T_n^2/T^2\right)\right)^2 + \left(2\varsigma(T_n/T)\right)^2}} \tag{9.13}$$

$$DAF = \frac{1}{1-(T_n/T_w)} \tag{9.14}$$

where
T = wave period
T_n = natural period of structure (first mode)
ς = damping factor (5% for steel structures in water)
T_w = wave frequency

Foundation Design

10

A foundation is part of the offshore platform and transmits the weight of the structure to the ground. The foundation should be designed to carry static, cyclic and transient loads without excessive deformations or vibrations in the platform.

The offshore platforms shall be fixed to the seabed by means of piles either driven through the main legs of the jacket or through skirt sleeves attached to the jacket legs or the combinations of both main and skirt piles.

In the design of a platform foundation, the following must be considered:

- Settlement characteristics of the soil
- Ultimate bearing pressure of the soil
- Stability of the soil
- Nature of foundation
- Nature of loads
- Allowable settlement
- Factor of safety

Loading of a soil produces settlement and can cause soil instability such as ground heave. In addition to imposed loads, the stability of a soil can be affected by environmental factors (e.g. displacement from sea current or seismic forces). The nature of the foundation also affects the distribution of stresses in the soil. Flexible and rigid foundations provide different load transfers to the soil and so do the eccentricity of loading, size and shape of foundation.

The ability of a soil to support a structure at a particular elevation for a specific period of time is of the upmost importance to the foundation engineer.

10.1 TYPES OF FOUNDATIONS

The two types of foundations commonly used for fixed offshore platforms are (SJR 1994; API 2000)

a. Pile foundation
b. Shallow foundation

They may be required for temporary and permanent use. The choice of a particular foundation type depends on the following:

- Nature and magnitude of loads to be carried
- Stability of the seabed
- Nature of soils at seabed level
- Slope of seabed
- Nature of sub-seabed strata

10.2 PILE FOUNDATION DESIGN

Piles are structural members made of timber, concrete or steel tubes embedded in the subsea bed strata that transmits the load of the superstructure to the lower layers of the soil. Steel tubes are commonly used as piles. Their load-carrying capacity is dependent on the following:

- Friction between the soil and the surface of the pile
- End bearing between the pile and soil
- Horizontal and vertical load-carrying capacity of soil
- Strength and stiffness of the pile

The most common methods of installing a pile are

- Driving the pile into the soil (usually used)
- Drilling a hole into the soil, lowering in the pilling and grouting the space between the soil and the pile

The types of pile foundations used to support fixed offshore platforms are

a. Driven piles
b. Drilled and grouted piles
c. Belled piles

Driven piles disturb the soil by remoulding the soil. It is possible however that the nature of soil may determine the method of piling. For example,

- A weak clay may require a bored pile
- Sand may be too unstable for unlined boring
- Chalk or hard coral may become too damaged by pile driving to support the pile

To develop skin resistance under load, a pile must slip. In particular, the slip of a pile may be such that the ultimate frictional resistance is reached at the top of the pile, whilst at the lower end, frictional and end-bearing resistance is less than the ultimate frictional resistance. The ultimate pile capacity is therefore not equivalent to the ultimate skin resistance.

Skin friction varies with depth and soil type. Therefore, the load transferred to the soil from the top to the bottom of the pile may not necessarily vary linearly.

It should be noted that if the soil plug in an open-ended pile is assumed to resist end-bearing loads, then the soil in the pile should be checked to see if it will develop sufficient pile wall friction to retain itself.

Each project must acquire a site-specific soil report showing the soil stratification and its characteristics for load bearing in tension and compression, shear resistance and load-deflection characteristics of vertically (axially) and laterally loaded piles. The soil borings at the desired location, and then performing in situ and laboratory tests, are necessary for developing data usable for the platform design. The soil report should show the calculated minimum axial capacities for piles of the same diameter as the platform design piles, soil resistance drive (SRD) curves, different types of mudmat bearing capacities, pile group action curves, shear resistance values, pile tip end-bearing values and lateral pile axial capacity values.

These values will be input into the structural analysis model (normally in StruCad, FASTRUDL or SACS software) and will determine minimum pile penetrations and sizes, considering a factor of safety of 1.5. For operating loads, the factor of safety must be 2.0 for piles. The unity check ratios must not exceed 1.0 in the piles or anywhere else in the platform in normal conditions.

The soil data required to assess the capacity of a statically loaded pile are similar to that required for shallow foundations (i.e. cohesion 'c', internal angle of friction θ and settlement characteristics).

Piles subjected to cyclic and lateral loadings, however, require the input of additional soil data such as elastic recovery from deformation, lateral modulus of subgrade reaction, lateral stress/strain modulus, and so forth.

The soil test and its interpretation should take into consideration the proposed method of piling.

Pile foundations are usually placed in groups, often with spacing 'S' of 3 to 3.5B, where 'B' is the pile diameter. Smaller spacings are often not desirable because of the potential for pile intersection and a reduction in load-carrying capacity. A pile cap is necessary to spread vertical and horizontal loads and any overturning moments to all of the piles in the group. The cap of onshore structures usually consists of reinforced concrete cast on the ground, unless the soil is expansive. Offshore caps are often fabricated from steel.

Refer to Section 6.3 of API RP 2A for detailed design of a pile foundation.

10.2.1 Vertically Loaded Piles

Methods of estimating end-bearing and skin friction capacity are based on the works of Terzaghi, Vesic, Hansen, Meyerhof, Tomlinson and others (US Army 1992; SJR 1994). Vertically loaded piles are also known as axially loaded piles (Figure 10.1).

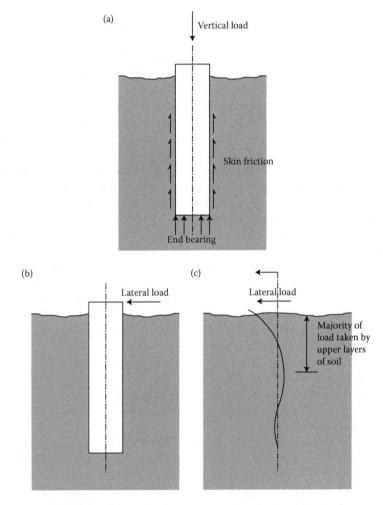

FIGURE 10.1 Pile loading and deflection. (a) Vertically loaded pile, (b) laterally loaded pile and (c) deflected form. (From SJR. 1994. *Subsea Structure Design.* London: JP Kenny Group.)

10.2.1.1 Ultimate Bearing Capacity

When the ultimate load applied on the top of the pile is Q_d, a part of the load is transmitted to the soil along the length of the pile, and the balance is transmitted to the pile base. The load transmitted to the soil along the length of the pile is called the ultimate friction load or skin load, Q_f, and that transmitted to the base is called the base or point load, Q_p. The total ultimate load, Q_d, is expressed as the sum of these two, that is, for axially/vertically loaded piles, the ultimate bearing capacity of a pile can be estimated as (API 2000)

$$Q_d = Q_f + Q_p = fA_s + qA_p \tag{10.1}$$

where

Q_f = skin friction resistance
Q_p = total end bearing
f = unit skin friction capacity
A_s = side surface area of pile
q = unit end-bearing capacity
A_p = gross end area of pile

10.2.1.2 General Theory for End-Bearing Capacity

The general equation for base resistance may be written as

$$Q_p = \left(cN_c + qN_q + \frac{1}{2}\gamma BN_\gamma \right) A_p \tag{10.2}$$

where

d = width or diameter of the shaft at base level/depth of footing
q = effective overburden pressure at the base level of the pile

$$q = \gamma \times d \tag{10.3}$$

A_p = base area of pile
c = cohesion/shear strength of soil
γ = effective/buoyant unit weight of soil
N_c, N_q, N_γ = bearing capacity factors which take into account the shape factor
B = minimum lateral dimensions of footing

10.2.1.2.1 Cohesionless Soils

Cohesionless soil is composed of granular or coarse-grained materials with visually detectable particle sizes and with little cohesion or adhesion between particles. These soils have little or no strength, particularly when dry, when unconfined and little or no cohesion when submerged.

For cohesionless soils, $c = 0$, and the term $(1/2)\gamma BN_\gamma$ becomes insignificant in comparison with the term qN_q for deep foundations (SJR 1994; API 2000). The general form of end-bearing capacity equation reduces to

$$Q_p = (qN_q +)A_p \tag{10.4}$$

10.2.1.2.2 Cohesive Soils

Cohesive soils are fine-grained materials consisting of silts, clays, and/or organic material. These soils exhibit low to high strength when unconfined and when air-dried depending on specific characteristics. Most cohesive soils are relatively impermeable compared with cohesionless soils.

For cohesive soils such as saturated clays, $\varphi = 0$, $N_q = 1$ and $N_\gamma = 0$, the end-bearing capacity of the pile is

$$Q_p = (cN_c + q)A_p \tag{10.5}$$

The net ultimate base load is

$$Q_p = (cN_c)A_p \tag{10.6}$$

All parameters are as defined above.

10.2.1.3 Skin Friction and End Bearing in Cohesive Soils

For cohesive soils, the Skin friction, f, can be calculated from the equation below.

$$f = \alpha c \tag{10.7}$$

where

α = a dimensionless factor
c = undrained shear strength of the soil at the point in question

Refer to API RP 2A for the computation of α.

10.2.1.4 Shaft Friction and End Bearing in Cohesionless Soils

For cohesionless soils, the shaft friction, f, can be calculated as

$$f = Kp_0 \tan \delta \tag{10.8}$$

where

K = coefficient of lateral earth pressure (ratio of horizontal to vertical normal effective stress)

p_0 = effective overburden pressure at the point in question

δ = friction angle between the soil and pile wall (may be selected from table in API RP 2A)

For open-ended pipe piles driven unplugged, $K = 0.8$.
For full displacement piles, plugged or closed end, $K = 1.0$.
For a pile end bearing in cohesionless soils, the unit end bearing, q, may be computed by the equation

$$q = p_0 N_q \tag{10.9}$$

where

p_0 = effective overburden pressure at the pile tip

N_q = dimensionless bearing capacity factor

10.2.2 Laterally Loaded Piles

When laterally loaded, a pile deflects and loads the soil horizontally. To assess the load-carrying capacity of the soil under this condition, the lateral bearing capacity of the soil is estimated and compared with the estimated pile/soil reactions and deflections.

The magnitude of the pile deflections (y) and soil reactions (P) are dependent on the relative stiffness of the soil and pile. The lateral soil-carrying capacity of a pile is mainly dependent on the strength nature of the upper layers of soil (SJR 1994).

10.2.2.1 Load Deflection of P–Y Curves

The data required to represent the stiffness of the soil layers are presented as a curve of soil reaction/soil force, P, and horizontal deflection, Y, termed as P–Y curves. This, coupled with the characteristics of the pile and load, can be used to compute the soil loadings and deflections. Typical P–Y curves for different soils can be obtained from API RP-2A.

For the purpose of analysis, the soil is categorised as sand, soft clay and stiff clays. This is because the soils react to loading in different manners. P–Y curves may be derived from the following relationship (SJR 1994; API 2000):

$$P = A \times P_u \times \tan h \left[\frac{k \times H}{A \times P_u} \times y \right] \qquad (10.10)$$

where

 A = factor to account for cyclic or static loading condition
 $A = 0.9$ for cyclic loading

For static loading,

$$A = \left(3 - 0.8 \frac{H}{D} \right) \geq 0.9 \qquad (10.11)$$

where

 P_u = ultimate bearing capacity at depth 'H'
 k = initial modulus of subgrade reaction
 y = lateral deflection, inches (m)
 H = depth, inches (m)

The following outlines methods of assessing the capacity of laterally loaded piles and is taken from API RP 2A.

10.2.2.2 Lateral Bearing Capacity for Sand

The ultimate lateral bearing capacity at a particular depth may be estimated from the lesser of the following equations.

$$P_u \text{ shallow} = (C_1 \times H + C_2 \times D) \times \gamma \times H \qquad (10.12)$$

$$P_u \text{ deep} = C_3 \times D \times \gamma \times H \qquad (10.13)$$

where

 P_u = ultimate resistance (force/unit length)
 γ = effective soil weight
 H = depth
 D = average pile diameter from surface to depth
 C_1, C_2, C_3 = coefficients relating to φ determined from API RP 2A graph

10.2.2.3 Lateral Bearing Capacity for Soft Clay

For static loads the ultimate unit lateral bearing capacity of soft clay, P_u is assumed to vary from $8c$ to $12c$ depending on depth (API 2000). The ultimate lateral bearing capacity at a particular depth may be estimated from Equations 10.14 and 10.15.

$$P_u = 3c + \gamma X + J\frac{cX}{D} \tag{10.14}$$

and

$$P_u = 9c \quad \text{for} \quad X \geq X_R \tag{10.15}$$

where
P_u = ultimate resistance
c = undrained shear strength for undisturbed clay soil samples
D = pile diameter
γ = effective unit weight of soil
J = dimensionless empirical constant values ranging from 0.25 to 0.5 depending on the clay and is determined from field test
X = depth below soil surface
X_R = depth below soil surface to bottom of reduced resistance zone

$$X_R = \frac{6D}{(\gamma D/c)+J} \tag{10.16}$$

For static loads case, P–Y curves may be derived from Table 10.1 for short-term static loading.

TABLE 10.1 Data for generating
P–Y curve

P/P_U	Y/Y_C
0.00	0.00
0.50	1.0
0.72	3.0
1.00	8.0
1.00	∞

Note: Where P = actual lateral resistance, psi (kPa), Y = actual lateral deflection, in (mm), $Y_c = 2.5\varepsilon_c D$ and ε_c = strain which occurs at one-half the maximum stress on laboratory undrained compression tests of undisturbed soil samples.

10.2.2.4 Lateral Bearing Capacity for Stiff Clay

For static loads, the ultimate unit lateral bearing capacity of stiff clay ($c > 1$ Tsf or 96 kPa), P_u, is assumed to vary from 8c to 12c. Refer to API RP 2A-WSD for the $P-Y$ curves of stiff clay.

10.3 SHALLOW FOUNDATION DESIGN

Shallow foundations are those foundations for which the depth of embedment is less than the minimum lateral dimension of the foundation element. Typical shallow foundations encountered in offshore structures are the mudmat and grouted mattress. In the design of shallow foundations, the ultimate bearing capacity and settlement of the structure must be considered (SJR 2000).

According to API RP 2A-WSD (2000), when designing a shallow foundation, the following need to be considered:

a. Stability, including failure due to overturning, bearing, sliding or combinations thereof.
b. Static foundation deformations, including possible damage to components of the structure and its foundation or attached facilities.
c. Dynamic foundation characteristics, including the influence of the foundation on structural response and the performance of the foundation itself under dynamic loading.
d. Hydraulic instability such as scour or piping due to wave pressures, including the potential for damage to the structure and for foundation instability.
e. Installation and removal, including penetration and pull-out of shear skirts or the foundation base itself and the effects of pressure buildup or drawdown of trapped water underneath the base.

The two common types of shallow foundations used offshore are

a. Grouted bags and cushions: Flexible skins inflated by the insertion of cement grout.
b. Mudmat: Flat steel plates reinforced with a grillage of steel members. The load is thus spread over the seabed contact area of the plate.

The load carrying capacity of shallow foundations is dependent on the strength of the foundation, the soil and seabed stability, and the mudmat plan dimensions. The shallow foundation of an offshore platform is designed as follows.

10.3.1 Ultimate Bearing Capacity

Ultimate bearing capacity is the theoretical maximum pressure which can be supported without failure; allowable bearing capacity is the ultimate bearing capacity divided by a factor of safety.

Vesic, Terzaghi, Hansen and Meyerhof have developed equations for estimating the ultimate bearing pressure. The following is equation developed by Terzaghi (SJR 1994).

The Terzaghi ultimate bearing capacity theory states that a foundation is shallow if its depth is less than or equal to its width. Later investigations, however, have suggested that foundations with a depth, measured from the ground surface, equal to three to four times their width may be defined as shallow foundations (API 2000).

Terzaghi's formula (1943) takes into account soil cohesion, soil friction, embedment, surcharge and self-weight. It is given below as

$$Q_{ult} = cN_cK_c + \gamma DN_q + 0.5\gamma BN_\gamma K_\gamma \qquad (10.17)$$

Although considered conservative, the Terzaghi equations are popular for their ease of use.

Typical Terzaghi values of N are

For $\varphi = 0$ $N_c = 5.7$ $N_q = 1.0$ $N_\gamma = 0$

For $\varphi = 20°$ $N_c = 17.7$ $N_q = 7.4$ $N_\gamma = 5.0$

For a foundation in which the load is vertical and the base is horizontal, K is only required to cater for the shape of the foundation.

Terzaghi shape factors, K, are listed in Table 10.2.

TABLE 10.2 Terzaghi shape factors

SHAPE FACTOR, K	STRIP FOOTING	CIRCULAR BASE	SQUARE BASE
K_c	1.0	1.3	1.3
K_γ	1.0	0.6	0.8

Source: SJR. 1994. *Subsea Structure Design*. London: JP Kenny Group.

Substituting the above shape factors into Equation 10.18:
For square base foundations,

$$Q_{ult} = 1.3cN_c + \gamma DN_q + 0.4\gamma BN_\gamma \tag{10.18}$$

For continuous foundations or strip footing,

$$Q_{ult} = cN_c + \gamma DN_q + 0.5\gamma BN_\gamma \tag{10.19}$$

For circular base foundations,

$$Q_{ult} = 1.3cN_c + \gamma DN_q + 0.3\gamma BN_\gamma \tag{10.20}$$

where

$$N_q = \frac{a^2}{2\cos^2(45+\varphi/2)}$$

$$a = e^{(0.75\pi-\varphi/2)\tan\varphi}$$

$$N_c = (N_q-1)\cdot\cot\varphi \quad \text{for} \quad \varphi>0$$

$$N_c = 5.14 \quad \text{for} \quad \varphi=0$$

$$N_q = \frac{\tan\varphi}{2}\left(\frac{N_{py}}{\cos^2\varphi}-1\right)$$

Q_{ult} = ultimate bearing capacity
c = cohesion/shear strength
γ = effective/buoyant weight of soil
D = depth of footing
B = width or the diameter of the foundation
N_c, N_q, N_γ = bearing capacity factors obtained from API RP 2A or set formula for φ
φ = the effective internal angle of friction
K_c, K_q, K_γ = factors that take into account

- Load inclination $\theta°$
- Footing shape (circular, square, rectangular, strip)
- Depth of embedment, d
- Inclination of base
- Inclination of ground

The values for c and φ should be obtained from a geotechnical soils report. These values are derived by the soil mechanic, through drained or undrained, shear or compression tests on soil samples. The value of c may be presented as the compressive or shear strength of the soil. Several values of φ may be given for a particular soil sample.

For layered soils, the bearing capacity of each layer should be checked. At the moment, there appears to be no method of accounting for base shapes other than strip, square or round. It is therefore suggested that other shapes of foundation be approximated to that of a circular or square foundation (SJR 1994).

10.3.1.1 Undrained Bearing Capacity ($\varphi = 0$)

With Reference to API RP 2A-WSD (2000), the design bearing capacity in undrained conditions is calculated as follows

$$Q = (cN_cK_c + \gamma D)A' \tag{10.21}$$

where

Q = maximum vertical load at failure
c = undrained shear strength of soil
N_c = a dimensionless constant, 5.14 for $\varphi = 0$
φ = undrained friction angle = 0
γ = total unit weight of soil
D = depth of embedment of foundation
K_c = correction factor which accounts for load inclination, footing shape, depth of embedment, inclination of base and inclination of the ground surface
A' = effective area of the foundation depending on the load eccentricity

Refer to API RP 2A for further information on shallow foundation design, specifically sliding considerations and drained bearing capacity.

10.3.2 Deformation

Deformation/settlement of a shallow foundation affects the structural integrity of the platform, its serviceability and its components.

Deformation or settlement of the foundation is classified as

a. Immediate/short-term deformation
b. Consolidation/long-term deformation

10.3.2.1 Short-Term Deformation

It takes place at the same time as the loading or within seven days. The immediate settlement of the soil because of applied loading can be estimated by considering stresses and strains in the soil (SJK 1994).

$$U_v = \int_0^{L0} \frac{\Delta q \, dz}{E_s} \tag{10.22}$$

where

q = soil stress at depth z
E_s = stress/strain modulus of soil (provided in the soil report)
$L0$ = length of soil influenced by load ($L0 = 5B$, B is the width of the foundation)
U_v = deformation over range of $z = 0$ to $L0$

For hard stratum, E_s of the hard layer is taken as $10 \times E_s$ of the soft layer.

10.3.2.1.1 Estimating Short-Term Deformation Using API RP 2A Method

Short-term vertical deformation for a circular rigid base is calculated as

$$U_v = \left(\frac{1-v}{4GR} \right) Q \tag{10.23}$$

Short-term horizontal deformation for a circular rigid base is calculated as

$$U_h = \left(\frac{7-8v}{32(1-v)GR} \right) H \tag{10.24}$$

Rocking

$$\theta_r = \left(\frac{3(1-v)}{8GR^3} \right) M \tag{10.25}$$

Torsion

$$\theta_t = \left(\frac{3}{16GR^3} \right) T \tag{10.26}$$

where

U_v = vertical displacement
U_h = horizontal displacement
Q = vertical load
H = horizontal load
θ_r = overturning rotations
θ_t = torsional rotations
M = overturning moment
T = torsional moment
G = elastic shear modulus of the soil

$$G = \frac{E_s}{2(1+v)} \qquad (10.27)$$

v = Poisson's ratio of the soil
R = radius of the base

The above equations can also be used for a square base of equal area. The depth of influence is assumed to be approximately $5 \times B$.

10.3.2.1.2   Theory of Elasticity Method for Estimating Short-Term Deformation

Short-term vertical deformation for a rectangular base is calculated as

$$U_v = qB\frac{(1-v^2)}{E_s}I_sI_f \qquad (10.28)$$

where

U_v = vertical displacement
q_z = intensity of contact pressure of foundation
B = least dimension of base
E_s = soil stress/strain modular
v = Poisson's ratio

$$I_s = I_1 + \frac{(1-2v)}{1-v}I_2 \qquad (10.29)$$

I_1, I_s, I_2, are influencing factors relating to the length and breadth of the foundation and the depth of influence obtainable from foundation analysis and design book by Joseph Bowles. The stress method or secant modular of elasticity method can also be used to estimate U_v (Bowles 1958; SJR 1994).

10.3.2.2 Long-Term Deformation

Long-term deformation (or long-term settlement) of a shallow foundation occurs as a result of consolidation of the soils under load. The amount of consolidation and long-term settlement often depends on the length of time the soil is loaded. Consolidation of soil can occur because of other factors such as vibration, changes in water content, and so on (SJR 1994).

Estimates of settlement/deformation are usually made for the life of the structure to be supported. Most time-dependent settlement take place over a period of 3–10 years.

$$U_v = \int_0^{L0} \frac{\Delta q\, dz}{E_s} \qquad (10.30)$$

For fine-grained saturated soils, time is of concern and therefore estimates of soil strain are related to times of drainage and pore pressure (SJR 1994). Long-term deformation or settlement is calculated using the void ratio (e) data as

$$U_v = \int_0^{L0} \frac{\Delta e}{h + e_0} \qquad (10.31)$$

where

U_v = vertical settlement
e_0 = initial voids ratio of the soil
Δe = change of voids ratio
h = layer thickness

Long-term settlement, U_v, for a soil layer under imposed vertical load can be obtained from the following equation which is presented in API RP 2A as

$$U_v = \frac{hC}{1 + e_0} \log_{10} \frac{q_0 + \Delta q}{q_0} \qquad (10.32)$$

where

U_v = vertical settlement
h = layer thickness
e_0 = initial void ratio of the soil
C = compression index of the soil over the load range considered

q_0 = initial effective vertical stress
Δq = added effective vertical stress

Compression characteristics of the soil are determined from one-dimensional consolidation tests.

10.3.3 Applied Bearing Pressure

To find the pressure distribution under a shallow foundation, the following relationship may be used (SJR 1994).

$$F = \frac{P}{A} \pm \frac{My}{I} \tag{10.33}$$

where
F = ground pressure
P = vertical load
M = moment on base
A = soil contact area
I = second moment of area of soil contact area
y = distance of point consideration to centre of bearing area

10.4 GRAVITY-BASED STRUCTURES

10.4.1 Resistance to Sliding

With gravity-based structures, reliance is placed on self-weight and the frictional interface between the structure and foundation. For sliding calculations, a safety factor of 1.5 should be used:

Design sliding force $\times 1.5 \geq$ Actual sliding resisting force $\tag{10.34}$

The behaviour of the foundation in resisting sliding depends on the soil type encountered and whether or not significant embedment of the lower chords of the structure will occur.

Typical values for the friction coefficient for structural members where embedment is insignificant (i.e. $D < 100$ mm) can be found in standard textbooks or guidelines. These can be used as lower bound values when other data are not available.

10.4.1.1 Sliding on Sand

In the absence of soil data for the area of seabed concerned, the Coulomb effect and friction coefficients can be used, providing insignificant embedment occurs by the member in the foundation (SJR 1994).

Where the soil characteristics are known, the friction coefficient (μ) can be determined as follows:

$$\mu = \tan \phi_p \tag{10.35}$$

$$\mu = \tan(1.1\phi_c) \tag{10.36}$$

where
ϕ_p = internal angle of friction determined from plate load tests
ϕ_c = internal angle of friction determined from triaxial tests

10.4.1.2 Penetration of Skirts

Resistance to sliding due to skirt penetration of foundation can be determined as follows:

$$\text{Resisting force, } F_R = k_p \rho_s d \tag{10.37}$$

where
k_p = coefficient of passive earth pressure
ρ_s = submerged density of soil
d = depth of skid penetration

Table 10.3 shows typical values of K_p for varying internal angles of friction (SJR 1994).

TABLE 10.3   Values of the coefficient of passive earth resistance (K_p)

	ANGLE OF SHEARING RESISTANCE Φ (DEGREES)								
	15°	18°	21°	24°	27°	30°	33°	36°	39°
K_p	1.70	1.89	2.11	2.37	2.66	3.00	3.40	3.86	4.40
Soil type	Silty clays and clayey silts			Sandy silts and silts		Sands and gravels			

Note: The values assume a horizontal plane surface in the soil providing passive resistance.

10.4.1.3 Sliding in Soft Clay

Where embedment in soft-clay foundations occurs (i.e. for tubular chord(s) of a structure), the ultimate resistance of the soil can be related to the embedment ratio.

For $D < B$,

$$P = N_{cl}C_aD \qquad (10.38)$$

For $D > B$,

$$P = N_{cl}C_aB \qquad (10.39)$$

where

P = ultimate resistance (force/unit length of structural member)
N_{cl} = lateral bearing capacity factor
C_a = average cohesion
D = depth of embedment from seabed surface to underside of member
B = width of footing

For sliding calculations, therefore,

$$F_{TC} < \frac{P}{1.5} \qquad (10.40)$$

where F_{TC} is due to the lateral overall allowable load on the structure because of hydrodynamic and/or other loadings.

Mudmat calculations may be carried out using simple bearing capacity theory from Terzaghi relating to the length and width of the mudmat.

Fire Resistance 11

Fire resistance is the ability of construction or its element to satisfy, for a stated period of time, load-bearing capacity, integrity and insulation when exposed to fire.

The offshore platform must be designed and constructed in such a way that, in the event of an outbreak of fire within the platform, the load-bearing capacity of the platform will continue to function until all occupants (operation and maintenance personnel) have escaped, or have been assisted to escape, from the platform and any fire-containment measures have been initiated.

The four methods used to assess and define the structural fire endurance or fire resistance of steel structural members are as follows (James 1988; Bea 1991).

a. Empirically derived correlations
b. Heat transfer analyses
c. Structural performance evaluation
d. Structural (mechanical fire) analysis

Using these approaches, the fire endurance of offshore structural components such as platform jackets and module support trusses (frames) can be analysed and predicted (Bea 1991).

For fire-resistant designs of offshore structures, the standard hydrocarbon fire curve should be used. This fire has a much faster rate of initial increase in temperature (see Figure 11.1).

Fire resistance of a steel member is a function of its mass, its geometry, the actions to which it is subjected, its structural support condition, fire protection measures adopted and the fire to which it is exposed. Design provisions to resist fire are briefly discussed below (see Table 11.1).

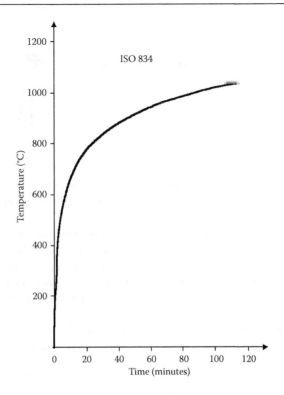

FIGURE 11.1 Standard time–temperature curve in a fire test. (Adapted from ESDEP. n.d. *ESDEP 4B.4.* Retrieved July 10, 2017, from http://fgg-web.fgg.uni-lj. si/%7E/pmoze/esdep/master/wg04b/t0400.htm)

11.1 EMPIRICALLY DERIVED CORRELATIONS

11.1.1 *W/D* Calculation Method

The *W/D* ratio is defined as the weight per unit length (*W*) of the steel member tested divided by the heated perimeter (*D*) of the member. Steel members with a lower *W/D* ratio than the referenced or tested steel member require an increased fireproofing thickness. On the other hand, steel members with a higher *W/D* ratio than the referenced steel member may have the sprayed fire-resistive materials (SFRM) thickness reduced.

TABLE 11.1 Empirical equations for steel columns

MEMBER / PROTECTION	SOLUTION	SYMBOL
Column/unprotected	$R = 10.3\left(\dfrac{W}{D}\right)^{0.7}$ for $\dfrac{W}{D} < 10$ $R = 8.3\left(\dfrac{W}{D}\right)^{0.8}$ for $\dfrac{W}{D} \geq 10$	R = fire-resistance time or time in minutes for the column to reach 1000°F W = linear density or weight of the steel section (lb/ft) D = heated perimeter of the steel section (in)
Column/gypsum wallboard	For critical temperatures of 1000°F $R = 130\left(\dfrac{hW'/D}{2}\right)^{0.75}$ where $W' = W + \left(\dfrac{50hD}{144}\right)$	h = thickness of protection (in) W' = weight of steel section and gypsum wallboard (lb/ft)
Column/spray-applied materials and board products, wide-flange shapes	$R = \left[C_1\left(\dfrac{W}{D}\right) + C_2\right]h$	C_1 and C_2 = constants for specific protection material

(Continued)

TABLE 11.1 (Continued) Empirical equations for steel columns

MEMBER / PROTECTION	SOLUTION	SYMBOL
Column/spray-applied materials and board products, hollow sections	$$R = C_1 \left(\frac{A}{P} \right) h + C_2$$	C_1 and C_2 = constants for specific protection material The A/P ratio of a circular pipe is determined by $$A/P_{pipe} = \frac{t(d-t)}{d}$$ where d = outer diameter of the pipe (in) t = wall thickness of the pipe (in) The A/P ratio of a rectangular or square tube is determined by $$A/P_{tube} = \frac{t(a+b-2t)}{a+b}$$ where a = outer width of the tube (in) b = outer length of the tube (in) t = wall thickness of the tube (in) (Continued)

TABLE 11.1 (Continued) Empirical equations for steel columns

MEMBER / PROTECTION	SOLUTION	SYMBOL
Column/concrete cover	$R = R_0(1 + 0.03m)$ where $R_0 = 10\left(\dfrac{W}{D}\right)^{0.7} + 17\left(\dfrac{h^{1.6}}{K_c^{0.2}}\right)$ $\times \left\{1 + 26\left[\dfrac{H}{P_c C_c h(L+h)}\right]^{0.8}\right\}$ $D = 2(b_f + d)$	$R_0 =$ fire resistance at zero moisture content of concrete (min) $m =$ equilibrium moisture content of concrete (% by volume) $b_f =$ width of flange (in) $d =$ depth of section (in) $K_c =$ thermal conductivity of concrete at ambient temperature (Btu/hrft°F) $h =$ thickness of concrete cover (in)
Column/concrete encased	For concrete-encased columns use $H = 0.11W + \dfrac{P_c C_c}{144}(b_f d - A_s)$ $D = 2(b_f + d)$ $L = (b_f + d)/2$	$H =$ thermal capacity of steel section at ambient temperature ($=0.11$ W Btu/ft°F) $c_c =$ specific heat of concrete at ambient temperature (Btu/lb°F) $L =$ inside dimension of one side of square concrete box protection (in) $A_s =$ cross-sectional area of steel column (in²)

Source: From James, M. A. 1988. Analytical methods for determining fire resistance of steel members. In *SFPE Handbook of Fire Engineering* (Chapter 6). Quincy, MA: Society of Fire Protection Engineers.

W/D, A/P and M/D ratios are used to size structural steel members with the objective of fire protection. Usually, as the W/D, A/P and M/D ratios increase, the fire resistance increases and/or the required thickness of directly applied fire protection material decreases for a given rating (Bea 1991).

$$\frac{W}{D} = \frac{M}{D} \times 0.017 \tag{11.1}$$

$$\frac{M}{D} = \frac{(W/D)}{0.017} \tag{11.2}$$

$$\frac{A}{P} = \frac{W}{D} \times \frac{144}{490} \tag{11.3}$$

where
   D = heated perimeter of steel section (in)
   A = cross-sectional area of steel section (in^2)
   P = heated perimeter of steel section (in)
   M = mass of steel section (kg/m)
   D = heated perimeter of steel section (m)

Structural steel columns should be insulated to avoid failure at temperatures of approximately 1000°F when exposed to fire. Laboratory experiments have revealed that, for a slenderness ratio from 42–112, the temperature at which the steel structure will fail can be approximately calculated by the equation below (Bea 1991).

$$[1040 + 1.8(1/r)] \pm 50F \tag{11.4}$$

The ensuing empirical relationship developed by Stanzak and Lie can be used to determine the fire endurance of steel columns (James 1988).

For $\dfrac{W}{D} < 10$

$$R = 10.3 \left(\frac{W}{D}\right)^{0.7} \tag{11.5}$$

And for $\dfrac{W}{D} > 10$

$$R = 8.3\left(\frac{W}{D}\right)^{0.8} \tag{11.6}$$

where

  R = fire resistance in minutes or time in minutes for the column to reach 1000°F
  W = linear density or weight of the steel section (lb/ft)
  D = heated perimeter of the steel section (in)

$$R = (C_1 W/D + C_2)h \tag{11.7}$$

where

  R = fire endurance or fire resistance (hr)
  W = steel weight per lineal foot (lb/ft)
  D = heated perimeter of the steel at the insulation interface (in)
  h = thickness of insulation (in)

The constants C_1 and C_2 need to be determined for each protection material. The constants take into account the thermal conductivity and heat capacity of the insulation material (James 1988).

The W/D ratio is used to express characteristic fire resistance of steel members. W/D ratio is used by laboratories for normalising structural steel fire-resistive ratings for fireproofed members. Steel members having a higher W/D ratio than the rated member size (for a given thickness of fireproofing) are considered larger than the specified minimum size required for achieving the preferred degree of fire resistance (AkzoNobel 2017; Bea 1991).

W/D ratio has a disadvantage of not accounting for geometry of the structural member or how it is used. W/D ratio is used for wide-flange sections and A/P for hollow sections.

The thickness for the protection of steel beams is determined based on the following equation:

$$h_1 = \left[\frac{(W_2/D_2)+0.6}{(W_1/D_1)+0.6}\right]h_2 \tag{11.8}$$

where

  h_1 = thickness of spray-applied fire protection (in)
  W = weight of steel beam (lb/ft)
  D = heated perimeter of the steel beam (in)

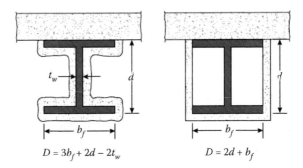

$$D = 3b_f + 2d - 2t_w \qquad\qquad D = 2d + b_f$$

FIGURE 11.2   Heated perimeter of steel beams. (a) Contour protection, (b) box protection. (From James, M. A. 1988. Analytical methods for determining fire resistance of steel members. In *SFPE Handbook of Fire Engineering* (Chapter 6). Quincy, MA: Society of Fire Protection Engineers.)

and where the subscripts
  1   = substitute beam and required protection thickness and
  h_2 = the beam and protection thickness specified in the referenced tested
      design or tested assembly.

Limitations of this equation are noted as follows.

a. $W/D \geq 0.37$
b. $h \geq 3/8$ in (9.5 mm)
c. The unrestrained beam rating in the referenced tested design or tested assembly is at least 1 hr.

The above equation only applies to the calculation of the protection thickness for a beam in a floor or roof assembly. All other features of the assembly, including the protection thickness for the deck, must remain unaltered (James 1988), (Figure 11.2).

11.1.2 *Hp/A* Calculation Method

The section factor (*Hp/A*) is the ratio of the fire-exposed perimeter to the cross-sectional area of the steel member. Essentially, it is a measure of how quickly the steel section will heat in a fire and, therefore, how much fire protection is required (AkzoNobel 2017; Bea 1991).

A fire-resistance test in accordance with the requirements of BS476, Part B, proves that, for a totally stressed unprotected steel section, columns exposed on four sides that have a section factor, Hp/A, of up to 50 m^{-1} can achieve a half-hour fire rating (Profire 2014; Bea 1991).

For a tubular member,

$$\text{Section Factor} = \frac{\text{Heat Perimeter}(H_p)}{\text{Cross} - \text{sectionalArea}(A)} = \frac{H_p}{A} \tag{11.9}$$

$$\frac{H_p}{A} = \frac{12.56(D)}{(D^2 - ID^2)} \tag{11.10}$$

where D is the outer diameter of the member, and ID is the internal diameter of the member.

11.1.2.1 Fire-Resistance Test

Fire testing involves live fire exposures upwards of 1100°C, depending on the fire-resistance rating and duration one is after. More items than just fire exposures are typically required to be tested to ensure the survivability of the system under realistic conditions.

A fire-resistance test should be performed in accordance with the requirements of ASTM E119, BS476, Part B, API RP 6F, ISO 5660-1, ISO 5657, API Std 607, BS 6755-2, DIN 53436, FTP Code, HSE (UK) Offshore safety reports, IEC 60331, IEC 60332-3, ISO 834, ISO 1182, ISO 1716, IMO res. A.653(16), IMO res. A.754(18) and client requirement.

Basic test standards for walls and floors are BS 476: Part 22: 1987, BS EN 1364-1: 1999 and BS EN 1364-2: 1999 or ASTM E119.

Walls, floors and electrical circuits are required to have a fire-resistance rating.

11.2 HEAT TRANSFER ANALYSIS

The purpose of the heat transfer analysis is to determine the time required for the structural member to attain a predetermined critical temperature or to provide input to a structural analysis. The temperature endpoint criteria cited by ASTM E119 are often accepted as the critical temperatures (James 1988).

Heat transfer analysis should be performed to define the steel temperature, taking into account the effects of radiation, convection and conduction. The offshore platform should be able to resist accidental and catastrophic fires.

The temperature distribution and relevant stresses may be evaluated by a simple finite element analysis (FEA) simulation (TMR 2009; Visser 1995). The modelling of a structure involves three stages:

a. The first stage is to model the fire scenario to determine the heat energy released from the fire and the resulting atmospheric temperatures within the platform.
b. The second stage is to model the heat transfer between the atmosphere and the structure. Heat transfer involves three phenomena (conduction, convection and radiation), all of which contribute to the rise in temperature of the structural materials during the fire event.
c. The third stage is the determination of the response of the structure— basic simple checks, engineering advanced models and sophisticated discrete models based on all data available.

Heat conduction, heat accumulation and exchange of radiation are calculated on the basis of the refined FEM model. The transient behaviour is based on the following equation.

$$KT + C\dot{T} = Q \qquad (11.11)$$

where

K = the conductivity matrix of the structure
C = the heat capacity matrix
T and $\dot{T}$ are the vector of nodal point temperatures and the corresponding rates, respectively
Q = the heat load vector

The above equation is solved in the time domain using a standard integration procedure. The temperatures in each structural member are transferred to the space frame model.

11.2.1 Heat Transfer Equation

The general approach to studying the increase of temperature in structural elements exposed to fire is based on the integration of the Fourier heat transfer equation for non-steady heat conduction inside the member (ESDEP 2017).

$$\frac{dQ_{cond}}{dx} = \frac{d}{dx}\left(\lambda_i \frac{dT}{dx}\right) = \rho c_p \frac{dT}{dt} \qquad (11.12)$$

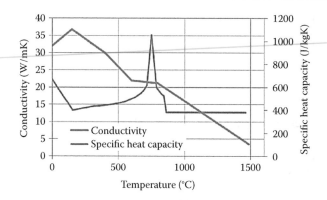

FIGURE 11.3 Thermal properties of steel at elevated temperature.

For a simplified calculation, use $c_p = 600$ J/kgK, $\lambda_i = 45$ W/mK, $\rho = 7850$ kg/m³ for all grades of steel (Figure 11.3).

11.2.1.1 Temperature Development of Steel Sections during Fire

$$\frac{dT}{dt} = \alpha \frac{d^2T}{dx^2} \tag{11.13}$$

$$\alpha = \lambda_i / \rho c_p \tag{11.14}$$

The quantity α is known as thermal diffusivity and varies with the temperature.

The quantity of heat transferred per unit length in the time interval Δt is

$$\Delta Q = \lambda_i A_m (T_f - T_s) \cdot \Delta t \tag{11.15}$$

where
λ_i = the total heat transfer coefficient (W/m²°C)
A_m = the perimeter surface area per unit length exposed to fire (m²/m)
T_f = the temperature of hot gases (°C)
T_s = the temperature of steel during the time interval Δt (°C)

If no loss of heat is considered, the internal energy of the unit length of a steel element increases by the same quantity ΔT_s.

$$\Delta Q = C_p \times \rho \times A \times \Delta T_s \qquad (11.16)$$

where
A = the cross-sectional area of the member (m²)

The temperature rise of the steel is given by combining the two equations above as follows

$$\Delta T_s = [\lambda_i / (C_p / \rho)] \times [A_m / A] \times (T_f - T_s) \Delta t \qquad (11.17)$$

Solving the above incremental equation step by step gives the temperature development of the steel element during the fire. In Eurocode 3 Part 1.2, it is suggested that

$$\Delta t \leq \frac{2.5 \times 10^4}{(A_m / A)} \qquad (11.18)$$

where
Δt is in seconds
A_m / A is in m⁻¹

There is a need to also calculate the critical temperature of structural steel members, although 1000°F is normally taken as the critical temperature by AISC. The critical temperature (θ_{cr}) which leads to the failure is calculated for a steel structure assuming a uniform temperature distribution along and across the members.
For unprotected elements, the equation is

$$\theta_{cr} = 1.85t \left(\frac{A_m}{A} \right)^{0.6} + 50 \qquad (11.19)$$

$$\frac{A_m}{A} = 0.36 \left[\frac{(\theta_{cr} - 50)}{t} \right]^{1.67} \qquad (11.20)$$

$$t = 0.54(\theta_{cr} - 50)(A_m / A)^{-0.6} \qquad (11.21)$$

Equations 11.19–11.21 are valid within following ranges: $t = 10$–80 min, $\theta_{cr} = 400°C$–$600°C$.

In the same way, for sections protected by a light insulation material, the equations are

$$t = 40(\theta_{cr} - 140) \times [dA/\lambda_i A_m]^{0.77} \tag{11.22}$$

$$d = 0.0083[t/(\theta_{cr} - 140)]^{1.3}[A_m/A]\lambda_i \tag{11.23}$$

where
   d = the protection thickness (m)
   λ_i = the thermal conductivity of the material (W/m°C)

All other parameters are as defined previously.

These equations can be expressed also in a nomogram, which is very practical for design purposes (see Figure 11.4).

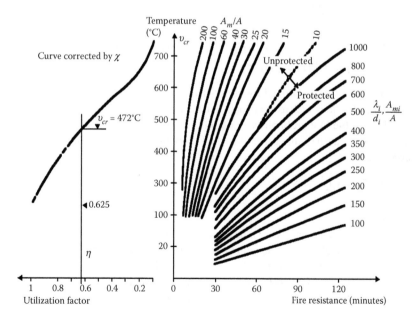

FIGURE 11.4   Relation between utilization factor, section factor and fire resistance. (From ESDEP. (n.d.). ESDEP WG 4B. Retrieved July 6, 2017, from http://fgg-web.fgg.uni-lj.si/%7E/pmoze/esdep/master/wg04b/t0100.htm)

11.3 STRUCTURAL (MECHANICAL FIRE) ANALYSIS

During fire, the load-bearing resistance of the steel decreases, leading to a decrease of mechanical properties such as yield stress, young modulus and ultimate compressive strength (Figure 11.5).

11.3.1 Applied Loads

The applied load is calculated as

$$P = P_u(\theta_{cr})　\text{(11.24)}$$

where

P = the applied load in fire conditions
θ_{cr} = the critical temperature
P_u = the load-bearing resistance at room temperature

11.3.1.1 Load Combination for Accidental Fire Situations

The applied load is obtained by considering the accidental combination of the mechanical actions such as dead load, live load, wind (only for bracing) and snow.

Due to the low probability that both fire and extreme severity of external actions occur at the same time, only the following accidental combinations are considered.

$$1.0G_k + y_1 Q_{k,1} + S y_{2,i} Q_{k,i}　\text{(11.25)}$$

where

G_K = the characteristic value of permanent actions (permanent/dead load)
$Q_{K,1}$ = the characteristic value of the main variable actions
$Q_{K,i}$ = the characteristic value of other variable actions
y_1 = the frequent value of the main variable actions
$y_{2,i}$ = the average of the other variable actions

Generally, in fire, $y_1 = 0.5$ and $y_{2,i} = 0$.

Apart from bracings, $Q_{K,1}$ and $Q_{K,2}$ generally correspond to imposed loads and snow loads (ESDEP 2017).

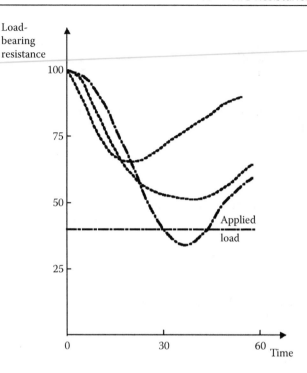

FIGURE 11.5 Examples of decrease of load-bearing resistance of a structure during exposure to fire. (From ESDEP. (n.d.). ESDEP 4B.3. Retrieved July 9, 2017, from http://fgg-web.fgg.uni-lj.si/%7E/pmoze/esdep/master/wg04b/t0300.htm)

11.3.2 Calculating the Load-Bearing Resistance of Steel Members

The critical temperature (θ_{cr}) leads to the failure and should be calculated for a steel structure assuming a uniform temperature distribution along and across the members.

11.3.2.1 Tension Members

At room temperature, the ultimate thermal resistance is given by

$$N_p = A \times f_y \tag{11.26}$$

where

N_p = ultimate thermal resistance
A = cross-sectional area of the member
f_y = yield stress

At a given uniform temperature q, through the member, the ultimate tensile resistance is

$$N_p(q) = A \times y(q) \times f_y \tag{11.27}$$

where $y(q)$ is the strength reduction of steel at q.

The collapse of the member will occur at the temperature θ_{cr} when

$$N_p(\theta_{cr}) = N \tag{11.28}$$

where N = the applied load in fire conditions.

Equation 11.28 can also be written as

$$A \times y(\theta_{cr}) \times f_y = A \times s \tag{11.29}$$

where s = applied stress in fire conditions.

Thus,

$$y(\theta_{cr}) = \frac{s}{f_y} \tag{11.30}$$

or

$$y(\theta_{cr}) = \frac{A \times s}{A \times f_y} = \frac{N}{N_p} = \frac{P}{P_u} \tag{11.31}$$

Therefore, knowing P/P_u makes it possible to determine the value of the steel critical temperature, θ_{cr}, for which $y(\theta_{cr})$ is equal to P/P_u, using Figure 11.6 (ESDEP 2017).

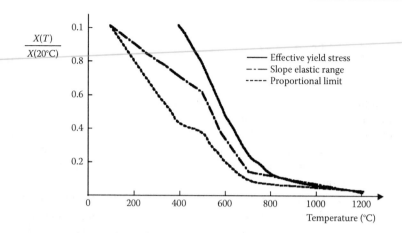

FIGURE 11.6 Parameters of structural steel at elevated temperature. (From ESDEP. (n.d.). ESDEP 4B.3. Retrieved July 9, 2017, from http://fgg-web.fgg.uni-lj. si/%7E/pmoze/esdep/master/wg04b/t0300.htm)

11.3.2.2 Columns

In the analysis of the columns, the effects of column buckling is considered by modifying the ultimate bearing resistance by buckling coefficient.

$$y(\theta_{cr}) = kP/P_u \qquad (11.32)$$

where k is the correction factor; $k = 1.2$.

The correction factor is used to compensate the choice of, f_y, which is related to the effective yield stress (the stress level at which the stress-strain relationship of steel tends to a yield plateau for a certain temperature) and not to the yield stress at 0.2% strain. Both P and P_u should be evaluated using the appropriate bucking coefficient.

$$\gamma(\theta) = \frac{1}{\varnothing(\theta)} + [\varnothing(\theta)^2 - \lambda(\theta)^2]^{(1/2)} \leq 1 \qquad (11.33)$$

where

$$\varnothing(\theta) = 0.5(1 + \alpha(\lambda(\theta) - 0.2) + \lambda(\theta)^2) \qquad (11.34)$$

and

$$\lambda(\theta) = \frac{\lambda}{\pi}\sqrt{\frac{f_y(\theta)}{E(\theta)}} \approx \sqrt{\frac{f_y}{E}} \tag{11.35}$$

In the case of a braced frame in which each storey comprises a separate fire compartment with sufficient fire resistance, the Eurocode 3 Part 1.2 recommends to use 0.5 as the effective buckling length coefficient (γ) of a steel column at high temperatures for columns when there are members of ordinary temperatures adjacent to both the upper and lower ends and to use 0.7 for columns on the top storey.

However, the Recommendation for Fire Resistant Design of Steel Structures by AIJ advises using 1.0 regardless of the boundary conditions of the column at both ends since elongation of adjacent beams may cause local buckling of the column at the upper and lower ends and loosen the rotational restriction of the column.

11.3.2.3 Simply Supported Beams

For a simply supported beam which is uniformly loaded, the maximum bending moment is

$$M = \frac{PL}{8} \tag{11.36}$$

and the corresponding maximum stress is

$$s = \frac{M}{S_e} \tag{11.37}$$

where S_e is the minimum elastic modulus of the section.

Failure will occur when the total load on the beam is

$$P_u = \frac{8M_u}{L} \tag{11.38}$$

where M_u = plastic bending moment resistance given by

$$M_u = z \times F_y \tag{11.39}$$

and z is the plastic modulus of the section.

When the temperature is equal to θ, this plastic bending moment resistance is equal to

$$M_u(\theta) = z \times y(\theta) \times F_y \tag{11.40}$$

For a beam subject to a load of P, the collapse will occur at θ_{cr} when

$$P_u(\theta_{cr}) = P \text{ or } M_u(\theta_{cr}) = M \tag{11.41}$$

That is, when

$$y(\theta_{cr}) = S_e \times \frac{s}{z} \times F_y = \frac{s}{((z/s_e) \times F_y)} = \frac{P}{P_u} \tag{11.42}$$

where $f = Z/S_e$ is the shape factor of the steel section ($\sim$1.10–1.3).

11.3.2.4 Continuous Beams

For a continuous beam, the maximum bending moment is

$$M = \frac{PL}{8} \tag{11.43}$$

In a fire situation, a plastic hinge will form at the middle support as the temperature increases when

$$M_u(\theta_1) = M \tag{11.44}$$

The load-bearing resistance of this continuous beam is

$$P_u(\theta_{cr}) = \frac{12 M_u(\theta_{cr})}{L} \tag{11.45}$$

$$y(\theta_{cr}) = 8 \times S_e \times \frac{s}{12} \times z \times F_y = \frac{P}{P_u} \tag{11.46}$$

The ratio $12/8 = 1.5 = c$ is the statically indeterminate coefficient, or plastic redistribution coefficient.

11.3.2.5 Beam Columns

When axial force and bending moment act together on the same structural element, its critical temperature can be obtained from the following formula

$$y\left(\theta_{cr}\right) = \frac{N}{c_{min}N_p} + \frac{k_y M_y}{M_p c_y} + \frac{k_z M_z}{M_p c_z} \tag{11.47}$$

where, c_{min} is the lesser of the buckling coefficients, c_y and c_z about the y_y or z_z axis and k_y and k_z are the reduction factors for the y_y and z_z axes, respectively.

11.3.2.6 Steel Elements with Non-Uniform Temperature Distribution

For non-uniform temperature distribution in the structure, the global coefficient called the Kappa factor accounts for the beneficial influence of thermal gradient for beams (ESDEP 2017). (Figure 11.7) For a beam, the general formula becomes

$$y(\theta_{cr}) = k \times \frac{s}{c} \times f \times F_y \tag{11.48}$$

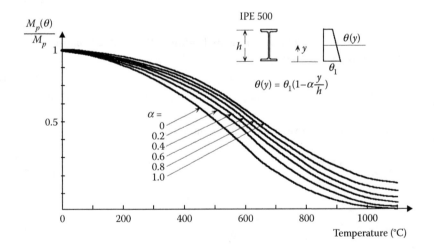

FIGURE 11.7   Effect of non-uniform temperatures on bending capacity. (From ESDEP. (n.d.). ESDEP 4B.3. Retrieved July 9, 2017, from http://fgg-web.fgg.uni-lj.si/%7E/pmoze/esdep/master/wg04b/t0300.htm)

where
  $k = 1$ for simply supported beams exposed to fire on all sides
  $k = 0.7$ for simply supported beams exposed on three sides
  $k = 0.85$ for continuous beams exposed on all sides
  $k = 0.60$ for continuous beams exposed on three sides

11.3.3 Load-Bearing Resistance of Composite Components

Structural components, members, and structural steel frame systems should be designed to maintain their load-bearing function and to satisfy other performance requirements specified for the client.

There is a need to calculate the load-bearing resistance of composite columns, composite slabs and composite members as a whole. Also, fire resistance of connections between columns and beams should be determined (ESDEP 2017).

11.3.3.1 Composite Beams

The tensile force summed over the three parts of the steel section is

$$T = \sum_{i=1}^{3} A_i y(\theta_i) F_y \tag{11.49}$$

where
  A_i = the area of lower flange, web and upper flange of the steel profile
  θ_i = the respective temperature

The point of application of this force is the plastic neutral axis at elevated temperature of the three parts of the steel action.

In order to balance this tensile force, a layer of the concrete slab is compressed such that

$$T = c = btf_{ck} \tag{11.50}$$

where
  b = the effective width of the slab
  t = the thickness of the compressive zone
  f_{ck} = the ultimate strength of concrete

The ultimate bending resistance of the composite section is

$$M^{+}_{u(\theta)} = T \times z \tag{11.51}$$

where z is the distance between the points of application of the tensile and the compressive forces.

For continuous beams, the determination of the full load-bearing resistance also requires the calculation of the negative plastic bending moment $\left(M^{-}_{u(\theta)}\right)$, assuming that stability in fire is maintained if the isostatic bending moment (M) of the applied load in a span is lower or equal to the sum of the positive and negative moment resistances of the composite section (ESDEP 2017):

$$M^{+}_{u(\theta)} + M^{-}_{u(\theta)} \geq M \tag{11.52}$$

11.3.3.2 Composite Slabs

The calculation of the fire behaviour of composite concrete slabs with profiled steel sheets is made using the same theories as for composite beams.

In this state, after 30 minutes of ISO fire, the steel sheet is not taken into account when calculating the mechanical behaviour of the element (ESDEP 2017).

11.4 STRUCTURAL PERFORMANCE EVALUATION

Design the offshore structure to have sufficient residual strength and structural robustness. Design the offshore structure for damage tolerance. The structural frame and foundation should be able to provide the strength and deformation capacity to withstand, as a system, the structural actions developed during the fire within the prescribed limits of deformation. The structural system shall be designed to sustain local damage with the structural system, remaining stable as a whole.

11.4.1 Deflection Criteria

Calculate the total deflection and rate of deflection for loaded and heated steel beams by considering elastic and plastic strain due to the applied load, thermal strain due to thermal expansion and creep strain (James 1988).

Compare the calculated deflection and rate of deflection with the Robertson-Ryan criteria. Deflection of unheated beams requires computer solutions.

11.4.1.1 Robertson-Ryan Criteria

Based on the examination of 50 fire test reports, Robertson and Ryan (1995) found that structural failure of floors and beams could be satisfactorily defined as the point at which

$$y_c \geq \frac{1}{800} \frac{L^2}{h} \tag{11.53}$$

$$\frac{dy_c}{dt} \geq \frac{1}{150} \frac{L^2}{h} \tag{11.54}$$

The temperature criterion of failure for truss or joist during fire exposure is defined as

$$T_k = \frac{(\Delta H/R)}{\ln[37.5(L_k/L)Z_k]} \tag{11.55}$$

The temperature criterion of structural failure for beam during fire exposure is defined as

$$T_{ac} = \frac{(\Delta H / R)}{\ln[((150 / \pi^2))Z_{ac}]} \tag{11.56}$$

where
- ΔH = activation energy of the creep
- y_c = midspan deflection resulting from fire exposure (in or cm)
- dy_c/dt = deflection rate (in/hr or cm/hr)
- L = span length (in or cm)
- h = distance between upper and lower extreme fibres of structural components (in or cm)

11.4.2 Critical Temperature

The critical temperature is defined as the temperature at which the material properties have decreased to the extent that the steel structural member is no longer capable of carrying a specified load or stress level (James 1988). Eurocode allows the use of the critical temperature method for tension members and restrained beams but not columns or unrestrained beams.

$$\theta_{cr} = 39.19 \ln \left[\frac{1}{0.9674 \mu_0^{3.833}} - 1 \right] + 482 \tag{11.57}$$

where

θ_{cT} = critical temperature
μ_0 = is the degree of utilisation, which is the ratio of the design loading in fire to the design loading resistance at ambient temperature

$$\mu_0 = \frac{E_{fi,d}}{R_{fi,d,o}} \tag{11.58}$$

$E_{fi,d}$ = the design effect of actions for the fire design situation as outlined in EC1–1-2 and
$R_{fi,d,o}$ = the design resistance of the isolated member as described in EC3-1-2 clause 4.2.3, but at a time $t=0$, that is, ambient temperature.

Default values for critical temperatures can be obtained from BS EN 1993-12:2005.

The critical steel temperature which must be protected against should be defined. For example, structural steel between 200°C and 750°C, vessels between 200°C and 350°C, or a 140°C temperature rise for divisions where the critical temperature requirement is much lower to protect personnel on the other side of the division or in a safety refuge.

11.4.2.1 Critical Stress Equations

Elastic design

$$\sigma_{yT} = \frac{1}{F_e} \frac{Z_e}{Z_p} \tag{11.59}$$

Plastic design

$$\frac{\sigma_{yT}}{\sigma_y} = \frac{1}{F_p} \tag{11.60}$$

where

σ_{yT} = critical yield stress at elevated temperature, T
σ_y = yield stress at ordinary room temperature
F_e = factor of safety, elastic design
F_p = factor of safety, plastic design
Z_e = elastic section modulus
Z_p = plastic section modulus

For $0 < T \leq 600°C$,

$$\sigma_{yT} = \left[1 + \frac{T}{900\ln(T/1750)}\right]\sigma_{y0} \tag{11.61}$$

$$E_T = \left[1 + \frac{T}{2000\ln(T/1100)}\right]E_0 \tag{11.62}$$

For $T > 600°C$,

$$\sigma_{yT} = \left[\frac{340 - 0.34T}{T - 240}\right]\sigma_{y0} \tag{11.63}$$

$$E_T = \left[\frac{690 - 0.69T}{T - 53.5}\right]E_0 \tag{11.64}$$

$$\alpha_T = (0.004T + 12)\times 10^{-6} \tag{11.65}$$

where

σ_{yT} = yield strength at temperature T (MPa) (psi)
σ_{y0} = yield strength at 20°C (68°F) (MPa) (psi)
E_T = modulus of elasticity at temperature, T (MPa) 9 psi)
E_0 = modulus of elasticity at 20°C (68°F) (MPa) (psi)
α_T = coefficient of thermal expansion at temperature T (m/m°C)
T = steel temperature (°C)

11.4.3 Critical Loads

The critical load is defined as the minimum applied load that will result in failure if the structural member is heated to a temperature, T. The critical load can be expressed as a point load or distributed load (James, 1988).

Critical load calculations can be performed using algebraic equations or a computer program.

11.4.4 Residual Strength

Residual strength is the load or force that a damaged object or material can still carry without failing. Material toughness, fracture size and geometry, as well as its orientation, all contribute to residual strength. Residual strength is a measure of the platform's ability to sustain damage without failure (Bea 1991).

Find the system members that are most critical to maintaining structural integrity. Detect potential weak links in the system and understand how progressive collapse may occur. Residual strength analysis may be carried out using the AFGROW software.

Steel jacket-type offshore platforms can realize considerably greater natural fire endurance solely through the application of fundamental design considerations. Specifically, the fire resistance of unprotected steel members, and the endurance of the overall system capacity, may be improved by limiting D/t and kL/r ratios of critical above-water braces to a maximum of 30 and 60, respectively, and optimizing structural configurations of the framing system using X-frame configurations to maximize residual strength (Bea 1991).

11.4.4.1 Failure Criteria

11.4.4.1.1   Ultimate Strength
The simplest failure criterion assumes that failure occurs at the ultimate (or yield) strength of the material. Thus, failure occurs when

$$\sigma_f = F_{tu} \tag{11.66}$$

where
  σ_f = the fracture stress
  F_{tu} = the ultimate strength

This criterion is applicable primarily to uncracked structures.

11.4.4.1.2   Fracture Toughness: Abrupt Fracture
Fracture toughness is the ability of a material (e.g. steel pipe) to resist stress and prevent cracks in the material from spreading.

Technically speaking, fracture toughness is the ability of a material to deform under increasing tensile stress in the presence of a defect or crack without exhibiting rapid and extensive fracture propagation. Materials that have high fracture toughness can absorb larger amounts of energy (i.e. can withstand higher pressures or levels of stress) before an existing crack spreads.

The fracture toughness is a measure of the material's resistance to unstable cracking. In a fractured structure, the stress intensity factor (K) correlates the local stresses in the region of the crack tip with crack geometry, structural geometry and the level of load on the structure. When the applied load level increases, the K value also increases and reaches a critical value, at which time the crack growth becomes unstable.

The Irwin's failure criterion states that abrupt fracture occurs when the crack-tip stress-intensity factor reaches or exceeds the fracture toughness of the material. The corresponding applied stress at failure is called the fracture strength. The failure occurs when

$$K \geq K_{cr} \tag{11.67}$$

where $K = \sigma\sqrt{\pi a}$ (11.68)

For plane stress

$$K_{cr} = \sqrt{EG_c} \tag{11.69}$$

For plane strain

$$K_{cr} = \sqrt{\frac{EG_c}{1-v^2}} \tag{11.70}$$

where
K_{cr} = the material's fracture toughness
v = Poisson's ratio
G = strain energy release rate
σ = applied stress
a = half the crack length
E = young modulus

11.4.5 Robustness

Robustness is defined according to EN 1991-1-1, Part 1–7, as the ability of a structure to withstand events such as fire, explosions, impact or the consequences of human error without being damaged to an extent disproportionate to the original cause. A robust structure has the ability to redistribute load when a load-carrying member experiences a loss of strength or stiffness and exhibits ductile rather than brittle global failure modes (Szarka 2015).

In Eurocode EN 1990:2002 (CEN 2002), the basic requirement to robustness is given in clause 2.1 4(P): 'A structure should be designed and executed in such a way that it will not be damaged by events such as explosion, impact, and the consequences of human errors, to an extent disproportionate to the original cause.'

A robust structure has the following characteristics:

a. Ability to resist lateral loading at all stages of construction and throughout the life of the structure.
b. Ability to absorb impacts due to accidental loading.
c. Ability to tolerate inaccuracies/uncertainties in the design and construction process as well as platform/building movements.
d. Ability for the structure to redistribute loads safely.

Problems with structural robustness often occur in relation to

a. Partially built structures or unusual situations such as structures within structures (where the need for lateral restraints is often ignored).
b. Uplift or another form of instability in which the restrained forces are finely balanced with the destabilising actions.
c. Difficulties in connecting different materials such as timber roof trusses on top of concrete walls.
d. Over reliance on a single element to carry a large portion of a structure, such as transfer beams that support a number of columns and/or other beams.
e. Long-span beams and cantilevers that are prone to instability without adequate lateral restraints.
f. Flexible structures that are prone to large deformation.

From this, we can conclude that structural robustness is a performance characteristic relating to, but not the same as, strength and resilience.

11.4.5.1 Robustness Can Be Measured Using the Following Three Approaches

a. Risk-based robustness index
b. Reliability-based robustness index, also known as probabilistic robustness index
c. Deterministic robustness indexes

Offshore platforms should be provided with adequate robustness levels in order to avoid progressive collapse. There must be adequate thermal robustness in the design to provide the time (endurance) required to accomplish critical tasks such as platform shutdown, firefighting response, disembarkation, and so on (Bea 1991).

11.4.5.2 Risk-Based Robustness Index

The risk-based robustness index is defined as

$$I_{rob} = \frac{R_{dir}}{R_{dir} + R_{ind}} \tag{11.71}$$

where I_{rob} is the robustness index, R_{dir} and R_{ind} are the direct and indirect risks, respectively.

The values of I_{rob} can be between 0 and 1, with larger values signifying larger robustness. I_{rob} can have values very close to 1.0 with relatively large direct and small indirect risks (Szarka 2015).

11.4.5.2.1 Reliability-Based Robustness Index
Reliability-based robustness index is also known as probabilistic robustness index.

11.4.5.2.2 Redundancy Index
Redundancy index (RI) is defined as

$$RI = \frac{P_{f(damaged)} - P_{f(intact)}}{P_{f(intact)}} \tag{11.72}$$

where $P_{f(damaged)}$ and $P_{f(intact)}$ are probability of failure for a damaged and intact system, respectively.

The values of RI can be between 0 and ∞, the smaller values meaning more robustness (i.e. $P_{f(damaged)}$ is not much higher than $P_{f(intact)}$).

11.4.5.2.3 Redundancy Factor
Redundancy factor is defined as

$$\beta_R = \frac{\beta_{intact}}{\beta_{intact} - \beta_{damaged}} \tag{11.73}$$

where β_R is the redundancy factor, β_{intact} and $\beta_{damaged}$ are the reliability indexes of the intact and damaged system, respectively (Szarka 2015).

The value of β_R changes between 1 and ∞, and the higher the redundancy factor (β_R), the more robust the structure is.

11.4.5.2.4   Reliability Index

Reliability index is defined as

$$\beta = \Phi(1 - P(F))^{-1} \tag{11.74}$$

where β is the reliability index, Φ is the cumulative normal distribution function and $P(F)$ is the failure probability.

11.4.5.3   Deterministic Robustness Indexes

11.4.5.3.1   Reserve Strength Ratio

The reserve strength ratio (RSR) was proposed by Faber et al. (2007) is defined in Equation 11.75.

$$RSR = \frac{R_c}{S_c} \tag{11.75}$$

where R_c and S_c are the base shear capacity and design value, respectively, in ultimate limit strength (ULS), with the R_c value coming from a pushover analysis. The value of RSR can be between 1 and ∞, where the bigger number denotes more unaccounted capacity (Szarka 2015).

11.4.5.3.2   Residual Influence Factor

A simple and practical measure of structural redundancy (and robustness) used in the offshore industry is based on the so-called RIF-value (residual influence factor), (ISO 19902: 2008). In order to measure the effect of damage (or loss of functionality) of structural member i, on the structural capacity, the so-called RIF-value can be defined as the damaged strength ratio. The RIF's values can be between 0 and 1, with larger values indicating larger robustness or larger redundancy (Szarka 2015).

To specify better the effect of losing one particular member (i) the RIF value (damaged strength ratio) is defined as

$$RIF_i = \frac{RSR_{fail,i}}{RSR_{intact}} \tag{11.76}$$

where $RSR_{fail,i}$ is the RSR value of the platform given that member 'i' has failed.

11.4.5.3.3 Stiffness-Based Robustness Measure

Robustness can also be measured by using the determinant of the static stiffness matrix of the structural system.

$$R_s = min \frac{\det K_j}{\det K_o} \tag{11.77}$$

where R_s is the stiffness-based robustness measure, K_0 is the stiffness matrix of the intact structure and K_j is the stiffness matrix of the structure with the given member removed.

Haberland et al. (2008) stated that this expression needs further normalisation in order for it to give a value between 0 and 1.

11.4.5.4 Energy-Based Structural Robustness Criterion

This is a new structural robustness index and is used to analyse the structure in terms of energy balance. André et al. (2015) give the general expression of the structural robustness index, IR, as

$$I_R(A_L|H) = \frac{\text{Damages up to unavailable collapse state for hazard } h}{\text{Damages up to collapse state for hazard } h} \tag{11.78}$$

$$I_R(A_L|H) = \frac{D_{uc} - D_{1st\,failure}}{D_c - D_{1st\,failure}} \text{ with} \begin{cases} 0 \le I_R \le 1 \\ D_c - D_{1st\,failure} \end{cases} = 0 \to I_R = 1 \tag{11.79}$$

where

A_L = the leading action

H = $\{h_1, h_2 \dots h \dots h_n\}$, a set of hazard scenarios (for example, a set of different actions with determined values applied in a given sequence)

$D_{1st\,failure}$ = damage energy of the structure when the 'first failure' state takes place for hazard scenario considered

D_{uc} = the damage energy corresponding to the state where collapse is unavoidable, the 'unavoidable collapse' state, for the hazard scenario considered

D_c = the damage energy corresponding to the collapse state for the hazard scenario considered

If the value of the structural robustness index (I_R) is equal to 1.0, then the structure is very robust. For the hazard scenario considered, if the structural robustness index is 0, the structure completely lacks optimisation in terms of structural robustness, or is less robust.

11.4.5.5 Ductility

Ductility is the ability of the material to endure and resist after yielding energy and thus allowing energy to be dissipated in a stable manner and stresses to be reallocated without substantial deterioration of the structure's performance. Mild steel is an example of a ductile material that can be bent and twisted without rupture. Use sections with low width-thickness ratios and adequate lateral bracing. High-strength steels are generally less ductile (lower elongations) and generally have a higher yield ratio. High-strength steels are generally undesirable for ductile elements. Provide connections that are stronger than members. Recognise that compression member buckling is nonductile (Engelhardt 2016).

The material ductility is attained by material strain-hardening and/or by material deformation capacity.

There is a need to design the offshore structure to have redundant elements appropriately placed, be able to withstand increased loadings and have the ability to redistribute loadings (i.e. ductility).

In ductility design, consider strain-rate sensitivity, strength and stiffness degradation, cyclic behaviour and low cyclic fatigue of materials and joints.

If ductile members are used to form a structure, the structure can undergo large deformations before failure. This is beneficial to the users of the structures, as in case of overloading, if the structure is to collapse, it will undergo large deformations before failure and thus provides warning to the occupants. This gives a notice to the occupants and provides sufficient time for taking preventive measures. This will reduce loss of life.

Ductility permits redistribution of internal stresses and forces, increases strength of members, connections and structures, permits design based on simple equilibrium models, results in more robust structures, provides warning of failure and permits the structure to survive severe earthquake loading.

11.4.5.6 Ductility Ratios

All structural steel has a minimum strain capacity of 17% at low strain rates. The offshore platform should have sufficient toughness against brittle fracture not to limit strain capacity significantly at the high strain rates associated with blast response (Nallayarasu 2013).

The strain limits for evaluating the effect of temperature may depend on the class of steel used for the design. The limiting strains for different classes of steel sections are specified in design codes.

It is important to know the shape and plastic hinges that form in order to allow the maximum strain to be calculated. Reduce the concept of strain limit to a limiting deformation (i.e. ductility ratio). Ductility ratio is defined as

$$\text{Ductility ratio} = \frac{\text{total deformation}}{\text{deflection at elastic limit}} \tag{11.80}$$

The deflection at elastic limit (γ_{el}) is the deflection at which bending behaviour can be assumed to change from elastic to plastic. Transition from elastic to plastic does not occur at a specific deflection. The following assumptions are made to define deflection at elastic limit (γ_{el}):

- Type of loading
- Beam fixity
- Shape of stress-strain curve
- Rate of loading and hence hinge formation

Ductility ratio is an important parameter for designing structures against explosion-induced forces. The ductility ratio (μ) of the steel member can also be defined as the ratio of the ultimate displacement at failure (Δu) to the displacement at the yield point (Δy).

$$\mu = \frac{\Delta_u}{\Delta_y} \tag{11.81}$$

The ductility reduction factor is calculated as

$$R_\mu = \frac{F_c}{F_y} \tag{11.82}$$

where
F_c = the ultimate base shear in the linear elastic behaviour
F_y = the ultimate base shear in the nonlinear elasto-plastic response

For the duration of hydrocarbon explosions on offshore installations, the ductility of joints is the key safety element which relates to structural performance as well as the level of damage. Primary members of the installation must not collapse and provide safe escape after the event; all main connections should not have yield strength much higher than expected, which can overload and prevent yielding of adjacent members. Therefore, in blast engineering where safety is the main concern, requisite of ductility for materials and the responses against the accidental loads are critical issues, which can only be achieved through a better understanding of fracture characteristics, both brittle and ductile (Ali 2007).

The ductility ratios currently being used in design of structural components are obtainable from design codes.

The structural member is considered to fail if the ductility ratio has exceeded 20, as given by TM 5 1300.

11.5 FIRE PARTITIONING

Topside and other parts of an offshore platform consist of load-bearing, non-load-bearing and partitioning components. The load-bearing components are to withstand service loads during fire; partitioning components are to prevent the spread of fire to adjoining spaces. Fire resistance is quantified by strength, integrity and insulation. Strength is the ability of a structural member, such as column, beam, load-bearing wall, slab, and so forth, to withstand the service loads during fire (refer to Sections 4.2.4.1, 4.2.4.2, 11.3.1 to 11.3.3 of this book). It applies to any load-bearing member. Integrity and insulation are prescribed for partitioning components of the platform, such as walls, doors (e.g. fire walls, fire-rated doors, fire-resisting ducts or dampers, etc.). The ability of a partitioning element to limit the rise of temperature on its unexposed side is termed as insulation; while the ability to prevent hot gases to reach the unexposed side through cracks/fissures is called integrity. Fire resistance is measured by the time to which a structural member satisfies all three criteria (i.e. strength, integrity and insulation) as applicable in fire; this is termed the fire rating of the member.

A partition is defined in British Standards as an 'internal, dividing, non-loadbearing, vertical construction'. In European (CEN) standards it is defined as a non-load-bearing wall, and EOTA European Technical Approval Guideline for partitions (ETAG 003), is entitled 'Internal partition kits for use as non-loadbearing walls'. A partition may be used for space division within the offshore platform, to separate areas of different floors, or used as an independent lining to an external wall.

Safety guidelines have been set out to ensure that the spread of fire remains limited during an incident. And, further, that the safety of those present is guaranteed as much as possible, that they can escape if necessary and that the fire can be fought. The necessary provisions can be included when the platform is designed and built. The guidelines are stated in the ISO and EN standards and in the MODU code for drilling platforms and the IMO FSC (Fire Safety Code) code for ships (Nutec 2008).

A platform can be divided into a number of fire compartments. A production platform and an accommodation platform can be 'separated' from each other in this way. The intention of the compartmentalisation is to limit the spread of fire as much as possible. Compartmentalisation is achieved by placing partitions which are classified as class A, class B, class C and class H. Partitions of different strengths can be placed depending on the degree of compartmentalisation. In class A, B and H, the number after the letter shows how long (in minutes) the partitions are fire resistant or fire retardant. After the stated number of minutes, the rise in temperature on the side not exposed to fire is so high that the fire will spread by means of fire transport. The fire will no longer be resisted/delayed by the partition. H-120 is used especially as a partition wall between the production platform and accommodations. This wall is resistant to the extreme heat of a liquid fire for 2 hours. The start or the spread of a fire on an offshore installation should be prevented in the earliest stage possible (Nutec 2008).

Use a fire-rated partition or fire-resisting partition for which the fire-resistance performance has been determined according to the appropriate British or European standards. Similarly, the reaction to fire performance of the exposed surfaces should also be determined by the appropriate fire test standards. It is important to determine the fire resistance and the reaction to fire performance of a partition. The fire resistance of non-load-bearing partitions is evaluated by

- BS 476: Part 20: 1987: 'Method for Determination of the Fire Resistance of Elements of Construction', which details the general principles of fire-resistance testing
- BS 476: Part 22: 1987: 'Methods for Determination of the Fire Resistance of Elements of Non-Loadbearing Elements of Construction', which details the procedures for testing partitions
- BS EN 1363-1: 'Fire Resistance Tests – Part 1 – General Requirements'
- BS EN 1364-1: 'Fire Resistance Tests for Non-Loadbearing Elements: Part 1 – Non-Loadbearing Walls'

These test methods measure two criteria of the partition's behaviour in the fire test: insulation and integrity.

A fire-rated partition will not allow hot gases to pass from the fire compartment to the surroundings by creating a structure (i.e. a compartment) which does not collapse and contains the fire for a given period of time. The fire resistance of such partition should range from 30–240 minutes (or more). In the event of fire, we have to ensure that surface materials are difficult to burn; the fire performance of boards and wall coverings that make up the outer faces of partitions should be subjected to the guidance of appropriate regulations.

Care must therefore be taken to ensure that there are no adverse effects in performance between the fire door and the partition. If the ductwork is not fire rated, it must be fitted with a fire-resistant damper(s). If the duct is fire rated, a fire penetration seal must be incorporated to both faces of the partition between the duct and the partition. Electrical cables need to be sealed into the partition by a fire penetration seal compatible with both cables and a partition. The fire penetration seal for pipes will need to be flexible to allow for structural movement and/or thermal expansion of the pipe(s). Some pipes may be hot; others may be cold, dependent on their use. Lagging is often applied to conserve heat, and any penetration sealant must be applied onto the pipe and not onto the lagging. The temperature of the pipe must be taken into account when choosing the penetration seal material. Strength and robustness classification should comply with BS 5234:1987.

11.6 PRACTICAL WAYS OF ACHIEVING FIRE RESISTANCE OF STEEL STRUCTURES

For bare steel members, the fire-resistance time can be increased by over-sizing the members (i.e. increasing the wall thickness), by maintaining the member size but using a higher strength steel, by utilising the restraining effects of connections or by a combination of these methods (see Figure 11.8).

For protected steel members, the thickness of the insulation must be such that the temperature of the steel at the required fire-resistance time (taking into account its section factor) does not exceed the critical (or limiting) temperature.

Fire-resistive coating or intumescent coating may be applied at an appropriate thickness. The required thickness of insulation for a structural steel member may be determined by using a nomogram which relates critical temperature, applied load, section factor and fire resistance (ESDEP 2017).

Intumescent fireproofing is a layer of paint which is applied along with the coating system on the structural steel members. The thickness of this intumescent coating is dependent on the steel section used. Intumescent coatings are paint-like substances which are inert at low temperatures but which provide insulation by swelling to provide a charred layer of low conductivity materials at temperatures of approximately 200°C–250°C. At these temperatures, the properties of steel will not be affected. Most intumescent coatings can traditionally provide up to 60–120 minutes of fire resistance economically (e.g. thick-film epoxy intumescent).

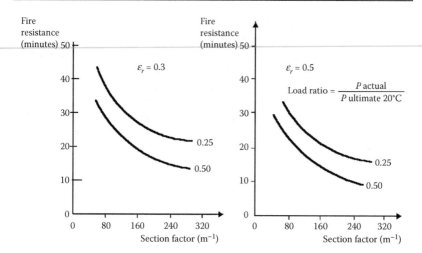

FIGURE 11.8   Fire resistance of bare steel beams as a function of section factor, different load levels and different resultant emissivity values. (From ESDEP. n.d.e *ESDEP WG 15A.1*. Retrieved July 5, 2017 from http://fgg-web.fgg.uni-lj.si/%7E/pmoze/esdep/master/wg15a/l0100.htm)

It should be noted that in the eventuality of a fire, the steel structure will collapse once the steel attains the critical core temperature (around 550°C or 850°F). The passive fire protection (PFP) system will only delay this by creating a layer of char between the steel and fire. Depending upon the requirement, PFP systems can provide fire ratings in excess of 120 minutes. PFP systems are highly recommended in infrastructure projects as they can save lives and property. PFP materials such as fire-resistant insulation products can be used either to envelop individual structural members or to form fire walls that contain or exclude fire from compartments, escape routes and safe areas.

Active fire protection (AFP) should be provided by water deluge and, in some instances, by fire-suppressing gas that is delivered to the site of the fire by dedicated equipment preinstalled for that purpose. Refer to API 2218, UL1709 and ISO 834.

For longer periods of fire resistance for floors with high loading and long spans, additional reinforcement may be necessary. Reinforce the walls with internal and/or external stiffeners. Fire- and blast-resistant walls should be used.

The fire resistance of tubular members can be improved by utilising the hollow interior to cool the load-bearing steelwork. Filling such members with water gives extremely high fire resistance when circulation is maintained.

A flexible/blanket system could be used. Concrete encasement also provides fire resistance. Spray-protection materials can also be used and some could be suitable for situations where the threat is from hydrocarbon fires. Fire doors should be encouraged for offshore applications.

Other areas of the topside that should be considered for fire resistance are nuts and bolts used in flanges (one of the weakest areas of any platform). Typical fire protection, which covers the complete flange, will not allow easy inspection of the units. By protecting only the nuts, regular inspection can be performed, reducing installation time and overall weight. Using moulded rubber-based material on just the flange nuts protects the stud bolts from elongating and the flange from breaking the seal during a fire.

Fire, Explosion and Blast Effect Analysis

12

The overall structural response of the fixed offshore platform to explosion loads may either be determined by either of the following two methods:

a. Nonlinear dynamic finite element analysis
b. Simple calculation models based on single degree of freedom (SDOF) analogies and elastic-plastic methods of analysis

Refer to Chapter 6.0 of DNV-RP-C204 for the design of offshore platforms against explosive loads. Also, in Chapter 18.0 of API RP 2A, an assessment procedure for fire and blast is shown in Figure 18.2-1.

12.1 SINGLE DEGREE OF FREEDOM ANALYSIS

A single degree of freedom (SDOF) system is a system whose motion is defined just by a single independent coordinate (or function). SDOF systems are often used as a very crude approximation for a generally much more complex system.

When the blast response of individual components or assemblages is characterised by a dominant deflection mode, simplified analysis based on an equivalent SDOF system can often be undertaken to evaluate blast resistance.

The SDOF approximation can provide an accurate assessment of the explosion resistance of individual components within a structure, the key requirements being the realistic representation of the boundary conditions, the strain-rate effect and the beam-column action. For the assessment of the overall structural response, the use of nonlinear finite element analysis has become essential, particularly for modelling complex interactions between the structural components (Izzuddin 1997).

12.1.1 Equation of Motion

The parameters for the simplified model are the mass m, stiffness or spring constant k, external force $F(t)$, structural resistance R and displacement y.

The blast load can also be idealised as a triangular pulse having a peak force F_m and positive phase duration t_d (see Figures 12.1 and 12.2). The forcing function is given as

$$F(t) = F_m\left(1 - \frac{t}{t_d}\right) \tag{12.1}$$

The blast impulse is approximated as the area under the force-time curve and is given by

$$I = \frac{1}{2}F_m t_d \tag{12.2}$$

The equation of motion of the undamped elastic SDOF system for a time ranging from 0 to the positive phase duration t_d is given by Biggs (1964) as

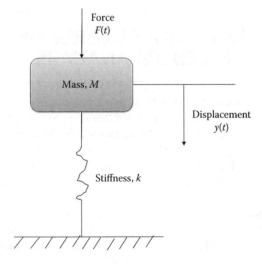

FIGURE 12.1   Single degree of freedom (SDOF) system.

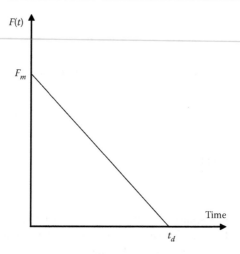

FIGURE 12.2   Blast loading.

$$my + ky = F(t) \tag{12.3}$$

$$my + ky = F_m\left(1 - \frac{t}{t_d}\right) \tag{12.4}$$

The general solution can be expressed as

Displacement,

$$y(t) = \frac{F_m}{k}(1 - \cos\omega t) + \frac{F_m}{kt_d}\left(\frac{\sin\omega t}{\omega} - t\right) \tag{12.5}$$

Velocity,

$$y(t) = \frac{dy}{dt} = \frac{F_m}{k}\left[\omega\sin\omega t + \frac{1}{t_d}(\cos\omega t - 1)\right] \tag{12.6}$$

In which ω is the natural circular frequency of vibration of the structure and T is the natural period of vibration of the structure which is calculated as

$$\omega = \frac{2\pi}{T} = \sqrt{\frac{k}{m}} \tag{12.7}$$

$$k = \frac{3EI}{L^3} \tag{12.8}$$

From the natural frequency, the natural period can be calculated as

$$T = \frac{2\pi}{\omega} \tag{12.9}$$

$$T = \frac{2\pi}{\omega} = 2\pi \left(\frac{m}{k}\right)^{0.5} \tag{12.10}$$

The maximum response is defined by the maximum dynamic deflection y_m which occurs at time t_m. The maximum dynamic deflection y_m can be evaluated by setting dy/dt in Equation 12.6 equal to 0, that is when the structural velocity is 0 (Ngo et al. 2007).

where
  ω = natural frequency of the SDOF system
  T = natural period
  t_d = duration or time taken for the overpressure to be dissipated
  m = actual mass (total mass)
  k = effective spring constant (or stiffness)

Duration 't_d' is related directly to the time taken for the overpressure to be dissipated (Ngo et al. 2007).

The energy dissipated by damping is very small in a system exposed to a very short pulse such as an explosion load (Chopra 2007). Thus, the damping effect is usually ignored when studying the structural response under a gas explosion load (Ki-Yeob et al. 2016).

12.1.2 The Dynamic Load Factor

The dynamic load factor (DLF) is defined as the ratio of the maximum dynamic deflection (y_m) to the static deflection (y_{st}) which would have resulted from the static application of the peak load F_m, which is shown as follows:

$$DLF = \frac{y_{\max}}{y_{st}} = \frac{y_{\max}}{F_m/k} = \varphi(\omega t_d) = \varphi\left(\frac{t_d}{T}\right) \tag{12.11}$$

The structural response to blast loading is considerably influenced by the ratio (t_d/T) or (ωt_d):

$$\frac{t_d}{T} = \frac{\omega t_d}{2\pi} \tag{12.12}$$

Three loading regimes are categorised as follows:

a. Impulsive loading regime

$$\omega t_d < 0.4 \tag{12.13}$$

b. Quasi-static loading regime

$$\omega t_d < 0.4 \tag{12.14}$$

c. Dynamic loading regime

$$0.4 < \omega t_d < 40 \tag{12.15}$$

The increase in the effect of a dynamic load is given by the dynamic amplification factor (DAF) or dynamic load factor (DLF). The DAF is equal to the DLF.

12.1.3 Hinge Rotation

Hinge rotation is another measure of member response which relates maximum deflection to span and indicates the degree of instability present in critical areas of the member. It is designated by the symbol θ (see Figure 12.3). Hinge rotation is also known as support rotation (ASCE 1997).

In Figure 12.3, θ_a is hinge rotation at support, and θ_b is the hinge rotation at centre $\approx 2\,\theta_a$.

12.2 ELASTIC-PLASTIC METHODS OF ANALYSIS

The nonlinear finite element approach is an accurate, even though computationally expensive, method for predicting the large displacement inelastic response

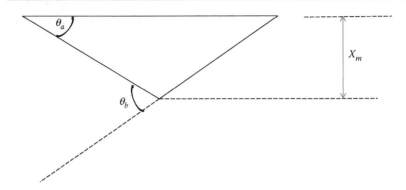

FIGURE 12.3   Hinge rotation.

of structures subject to explosion loading. Computational fluid dynamics (CFD) analysis is a very useful method that is able to model mechanical phenomena in complicated flow geometries (Ki-Yeob et al. 2016). The Flame Acceleration Simulator (FLACS) is a general tool used in the oil and gas industries; it is used to develop the potential overpressure caused by explosions in offshore facilities (Qiao and Zhang 2010).

12.3 EFFECT OF GAS EXPLOSIONS ON STRUCTURES

Most analysis has been based on explosives. Although peak pressure magnitudes may be similar, the very different pressure/time profiles of these and of gas explosions result in different levels and types of damage.

12.3.1 Duration of the Load and Natural Period of Vibration

It might be expected that the damage produced in a confined explosion would depend simply on the relative magnitude of the peak pressure generated and the pressure required to fail the confining structure. In reality, it is more complex. An explosion produces a pressure loading which varies with time, and the response of the structure or structural component is time dependent.

Structure response depends on

$$\frac{t_d}{T} = \frac{\text{Duration of imposed load}}{\text{Natural period of vibration}} \qquad (12.16)$$

that is, the ratio between the duration of the imposed load and the natural period of vibration of the structure.

There are three basic types of response:

i. $t_d > T$. Here the loading experienced will effectively be equivalent to a static load equal to the peak explosion overpressure.

ii. $t_d \approx T$. Here the loading experienced will effectively be equivalent to a static loading of a magnitude greater than the peak overpressure. The equivalent static overpressure can be up to $\pi/2$ times the incident overpressure.

iii. $t_d < T$. Here the pressure is effectively partially absorbed and the loading experienced will be equivalent to a static loading lower than the explosion peak overpressure, that is, a structure can withstand a higher dynamic pressure than static load necessary to cause failure.

Gas explosions will produce structural response type (i), that is $(t_d > T)$.

Thus, gas explosions can be considered approximately as static loadings equal to the peak overpressure.

Fire Protection Materials
13

13.1 TYPES OF FIRE PROTECTION MATERIALS

There are two basic types of fire protection: active and passive.

13.1.1 Active Fire Protection Materials

Active fire protection includes Fire protection systems which, in event of fire, can function only after its operation has been manually or automatically activated (ASFP 2015). Active fire protection consists of

a. The means to deliver the large volume of water required to tank cooling and firefighting purposes.
b. Alarms and fire and gas detection systems within all process areas of the platform, including living quarters.
c. Firefighting equipment.
d. Dry powder and sometimes foam fire suppression equipment or extinguishers, automatic fire sprinklers and water deluge systems. Refer to API 2218, UL1709 and ISO 834.
e. Effective communications systems so that the assistance of the emergency services can be quickly brought in.

13.1.2 Passive Fire Protection Materials

A fire protection system that performs its functions without the need for manual or automatic start of its operation in the event of fire is a passive system (ASFP 2015).

Passive fire protection materials protect steel structures from the effects of the high temperatures that may be produced in fire. They can be divided into two types: nonreactive, of which the most common types are boards and sprays, and reactive, of which thick-film intumescent coatings are the most common example. The intumescent coatings can be applied either on-site or off-site. The techniques used to protect steel are

a. Fire-resisting boards
b. Vermiculite cement sprays
c. Fibre sprays
d. Dry linings
e. Mineral wool
f. Intumescent coatings

13.1.2.1 Fire-Resisting Boards

These are fixed around the steel columns and beams to form a box and are commonly secured by fixing 'noggins' into the web of the steel with friction or adhesive and then securing the board to the noggin with screws. Based on individual formulation of the fire-resisting boards, they may serve as heatproof insulation for the steel. Properly fitted boards must have noggins of the same size, placed at the same intervals, as the original test installation and should have the same number of screws and other fixings (ASFP 2015).

The boards can be composed of gypsum-based plasters or calcium silicate, through fibres and specialist vermiculite-containing materials. There are varieties of thicknesses, and it is important to make sure that the correct board thickness is selected to give the requisite period of protection (ASFP 2015).

The boards are sometimes fixed with special, corkscrew-like screws, which are used when boards are of a mostly soft, pliable nature. In recent systems, some manufacturers have special spring clips to secure the bottom board; in other cases they rely on forming an almost self-supporting box. In many cases the boards will need decoration if they are exposed to view.

Detailed guidance on the installation of board protection systems is available from the Association for Specialist Fire Protection.

13.1.2.2 Mineral Wool

It is common to use mineral wool slabs or blankets to fully clad ceiling voids or other areas not requiring an aesthetic finish. In many instances, the mineral wool will have foil facing. The material is usually fixed with adhesives sticking 'noggins' into the web of the steel. Once again, it can easily be removed when surfaces are repaired and altered but will require refitting afterwards. Weld pins can also be used for fixing with large washers or clips to secure the blanket. This provides fire protection by wrapping a noncombustible layer of insulating material around the steelwork and thus slowing the rate of heat ingress into the steelwork because of the insulation (ASFP 2015).

13.1.2.3 Intumescent Paint Coatings

Thick-film intumescent coatings were originally developed for the offshore and hydrocarbon industries. Thick-film intumescent coatings are usually epoxy based and typically have a much higher dry film thickness than thin-film alternatives. Expansion ratios for thick-film intumescents are much lower than for thin-film materials, typically about 5:1. Aesthetic finishes are possible, and they can also be supplied in the form of preformed casings. Thick-film intumescent coatings can also be applied off-site.

The approval and use of intumescent coatings increased intensely in Europe in the 1970s as the major oil companies learned of their ability to protect structural steel from the dangerous heat caused by hydrocarbon fires, including jet fires caused by leaking hydrocarbons.

In 1988, an explosion and successive oil and gas fires at the Piper Alpha, a North Sea oil production platform, resulted in the deaths of 167 people. The brutality of this disaster, considered the worst offshore oil disaster at the time, provoked improved development and use of intumescent coatings for protection against hydrocarbon fires. The coatings developed tended to be thick-film coatings, often with mesh reinforcement.

13.1.2.3.1 How Do Intumescent Coatings Work?

These paints, or coatings, offer the most complex method of providing structural fire protection. They work by changing their nature from a decorative paint, which has been applied to the steelwork, into an intumescent layer of carbonaceous char, which forms when the coating is subjected to heat. This layer of char can be 50 times the thickness of the initial coat, and is formed as the paint is heated to around 200°C and above. At these higher temperatures the resin system melts and allows the release of a mineral acid, which reacts with a carbon rich element in the paint to form a carbon char (ASFP 2015).

Also released at the same time is a spumific, which provides a gas, which expands the foam to form the thicker layer. As the fire progresses and time passes, this layer of char grows thicker, thus increasing the insulation provided. The outer surface of it finally becomes soft and friable rather in the same way as the charcoal on a barbecue turns into a dry white powder.

During this period, energy is being absorbed by the reaction within the insulating layer being formed, thus limiting the amount of heat which passes through the coating to the steel below. The resin systems used to bind the chemicals together as paints vary and may use either a hydrocarbon solvent (usually xylene) or water as the dillulant of the resin. These have become known in industry jargon as either 'water-based' or 'solvent-based' intumescent, which is not technically correct but understood (ASFP 2015).

The materials are applied at very widely varying film thicknesses from as little as 275 μm for 30-minute protection in some areas up to 6 mm for 120-minute protection. This thickness is required from the base coat, which is the active element of the system and may require application in many coats at the higher thicknesses. They are most widely used for 30-minute and 60-minute work, where one spray coat will provide the required dry film thickness (DFT). A sealer coat is normally required and steel again should be blast cleaned and primed with a zinc phosphate primer in most cases. Many primer types can be used according to manufacturer instructions.

There are also materials that are known as 'thick-film intumescent', which are used mainly in the petrochemical industry. They work in the same way but differ by using epoxy resin systems rather than the softer types used in thin-film materials. The material expands at slower rate and to only five times the original. The char formed is much tougher, and the protection provided is capable of withstanding the high temperatures of hydrocarbon fires and jet fires. These products are rarely used in the building industry but common for steel protection in the offshore and petrochemical industry. They can be very successful on gas storage tanks and similar circumstances.

The use of intumescent paints requires blast cleaning of the steel and suitable primers and topcoats.

Manufacturer's literature will specify any special requirements, but in general, a zinc phosphate primer is used. The primer can be alkyd based for thin-film materials but must be epoxy based for the thick-film types. Primers containing metallic zinc must be thoroughly cleaned to remove any zinc salts and may not be suitable at all for water-based materials. In many cases, such primers will require a sealer coat, but this will be made clear in the manufacturer's instructions (ASFP 2015).

Most intumescent-coating suppliers provide guidance in calculating the thickness of the coating required for a specific use, and some have dedicated departments staffed with trained fire engineers who will do the calculations for you.

Intumescents can protect a variety of steel surfaces from structural columns and cellular beams, to building components, vessels and complex shapes. They can be formulated to protect against cellulosic and hydrocarbon fires, including jet fires and fires resulting from explosions.

13.1.2.4 Vermiculite Cement Sprays

These are undoubtedly the least expensive form of fire protection and can often be seen in places where a very rough and thick coating has obviously been applied to the profile of the steel. The material is supplied to the contractor as a dry powder made up of cement and exfoliated vermiculite. The vermiculite contains air, and protection is provided by a combination of water evaporation and insulation.

The material is mixed in a portable paddle mixer with water before application, and the amount of water is critical to the application and fire performance of the coating. It is then sprayed onto the surface using a screw feed pump usually manufactured by the company Putzmeister. Vermiculite sprays are widely used internally and also on petrochemical plants where special weatherproof grades are available. In certain circumstances, they require steel wire mesh support and weld pins to anchor the mesh. A water-repellent coat is often applied to give protection against moisture (ASFP 2015).

Thickness varies from around 10/12 mm up to 50 mm and, as the material is highly alkali, they require three epoxy or other alkali-resisting primers to be applied to the steel after blast cleaning.

This type of material can also be used for concrete protection and is predominantly common in France and certain other countries (ASFP 2015).

13.1.2.5 Fibre Sprays

These are very similar in nature to vermiculite sprays and everything detailed for vermiculite applies to these as well. The only difference is that this type of material uses mineral wool and insulates by providing a thick layer of heat-resistant material onto the surface (ASFP 2015).

13.1.2.6 Dry Lining Systems

Examples of dry lining systems are steel cladding systems, which are insulated with layers of mineral wool behind the steel cladding and other types of boarding systems built to create a void around the steelwork.

Such systems are often pre-finished and particularly popular when the level of aesthetic finish is important (ASFP 2015).

Fire Safety 14

Safety includes knowing what is dangerous and how to avoid the careless actions and inactions that make the area unsafe. Some examples of possible hazards that may cause explosions include

- Improperly grounded electrical equipment or wiring
- Fires from defective batteries
- Defective ventilation systems that cause a backup of hazardous fumes
- Improperly stored fuel or other combustible substances
- Poorly maintained hoses, pipes or pipelines that leak or spray fuel
- Improper use of welding devices or torches near combustible substances
- Corrosion of electrical equipment
- Failed equipment
- Failure to follow proper procedures for maintaining or cleaning the equipment
- Human error, for example, cutting into an operating pipeline with a torch
- Failure of a flange gasket or pump seal
- Failure to adhere to the Job Safety Analysis (JSA) and other safety procedures

14.1 SAFETY REGULATIONS

Employees working offshore must abide by the regulations that have been set out by U.S. Occupational Safety and Health Administration (OSHA), Bureau of Ocean Energy Management, Regulation and Enforcement. These standards have been developed to protect workers that are vulnerable to a whole new set of hazards when working out on the water, including regulations such as

- Maintenance must be done regularly on all equipment to ensure proper working condition
- Workers must receive training on how to implement proper safety techniques
- Oil rig owners must provide free medical care to all employees
- Equipment should be stored in a secure area that does not pose a risk of falling
- Weekly 'musters', or emergency exercises, should be carried out to prepare workers for real-life accidents
- Workers must have access to the General Platform Alarm (GPA) and/or emergency contact numbers

The Offshore Installations Prevention of Fire and Explosion, and Emergency Response Regulations 1995, highlight specific safety guidelines for fire prevention offshore. Some of these measures include

- Ensuring the appropriate handling, transportation and use of flammable/explosive materials (trained and informed personnel)
- Safely and securely storing flammable/explosive materials (away from sources of ignition, no chance of leakages)
- Trying to avoid accumulation of large quantities of such materials; intuitively the more flammable material stored on a facility the greater the risk of a fire and also its severity should one occur
- Not using such hazardous materials unless it is absolutely necessary

Regular safety inspections and audits should be encouraged.

The structural engineer should collaborate with a facility engineer knowledgeable in carrying out hazard analyses as defined in API Recommended Practice14J and with the operator's safety management system as described in API Recommended Practice 75.

Diesel fuel storage tanks should be isolated as far as practicable from ignition sources. Diesel fuel tanks should be enclosed by curbs, drip pans or catchments, which drain to a sump with provisions to prevent vapour return. Diesel fuel tanks should be adequately vented or equipped with a pressure/vacuum relief valve and should be electrically grounded. Fire-detection devices such as fusible plugs should be installed in the diesel fuel storage area.

Escape routes and safe areas should be provided and maintained to allow sufficient time for platform evacuation and emergency response procedures to be implemented during a fire or blast.

The industry has set up and continually improves training for safety awareness and emergency response. Every employee should attend a basic firefighting course as part of the BOSIET (Basic Offshore Survival Induction and Emergency Training) course.

Ignition sources should be minimised, and occupants and operators of the facility should be educated concerning operation and maintenance of fire-related systems for correct function and emergency procedures including notification for fire service response and emergency evacuation. By avoiding hot work onboard, fire hazards and shutdown requirements can be reduced.

Upon arrival on the offshore platform, there is a need for safety briefing on the layout of the platform, emergency response procedures, safety rules and required personal protective equipment, alarm signals, permit to work system, muster point and good housekeeping rules. Familiarise yourself with the platform, and take a safety tour of the facility (Nutec 2008).

Personnel are encouraged to wear personal protective equipment such as fire retardant overalls, safety helmet, gloves, safety boots and safety glasses (Risanto 2015).

There is a need to conduct safety drills on a regular basis by sounding an alarm to see the response according to the muster list.

Daily tool box meetings and weekly safety meetings should be encouraged. Safety personnel shall discuss safety-related issues. Employees are encouraged to relax after a hard day's work. There is a need for a dedicated smoking area and smoking policy. Workers should refrain from smoking in bed and restricted areas. Every country has a governing safety at work rules and there is a need to adhere to such rules and regulations.

Emergency shutdown systems such as a blowout preventors (BOPs) can be used to automatically shut down the wells when operating above allowable operating pressure. Install a Christmas tree on the platform where there is a need to control the amount of oil/gas entering the platform from the well.

A gaseous fire suppression system or fixed foam pourer system may be used. Install detectors for gases, smoke and fire; also, train personnel to deal with gas alarms. Keep a 500 m safety zone around the platform to minimise the chances of its collision by a ship. Encourage protection by blast walls and sprinkler systems (Profire 2014; Bea 1991). Barriers constructed from fire-resistant materials will be helpful in special situations to prevent the spreading of flames and to provide a heat shield. For example, in the Piper Alpha platform disaster, the fire walls were 'fire walls' only—they were not designed to withstand gas explosions.

Passive fire protection (PFP) materials, such as fire-resistant insulation products, can be used either to envelop individual structural members or to form fire walls that contain or exclude fire from compartments, escape routes and safe areas.

Active fire protection (AFP) should be provided by water deluge and, in some instances, by fire-suppressing gas that is delivered to the site of the fire by dedicated equipment preinstalled for that purpose (refer to API RP 2A-WSD).

Use work permits for work with chemicals and enforce safe working procedures as well as wearing the required personal protective equipment (PPE) such as air mask. Control hot work process, and if the need be, use appropriate Lock Out and Tag Out to ensure that no untrained personnel tampers with any process facility.

Provide fixed and portable radio communication on the offshore platform. The offshore platform should be equipped with commercial satellite communication or Global Maritime Distress Safety System (GMDSS) equipped with the required transmitters and receivers. Emergency Position Indication Radio Beacons are required on ships and offshore installations.

Establish a Search and Rescue (SAR) organisation and provide them with an SAR plan and the means for carrying it into effect (Nutec 2008).

Helicopter transport to the platform is safest. On the other hand, supply boat and crew basket or swing rope are more dangerous and, depending on the weather, it can be suspended. Helicopter decks are to be of steel or equivalent fire-resistant construction. If the space below the helicopter deck is a high fire-risk space, the deck is to be an 'A-60' Class division as given in Lloyd's PT 8 Chapter 1.

There is a need to train personnel on basic first aid and have a first aid and medical assistant on board the platform. Provide sick bays with the necessary medical equipment. Medical evacuation may be necessary where a casualty or a sick person is in a condition that medical assistance from shore is needed. There is a need for periodical medical training.

Put fire extinguishers at vantage points of the offshore platform, especially at exits. There is a need to train personnel on offshore firefighting, how to use fire extinguishers and have a firefighting team on board the platform always. There is a need to practise or attend every seven days: the alarm procedure, communication, work procedure, instructions about the various types of fire and use of fire equipment.

Firefighting systems and equipment vary depending on the age, size, use and type of structure. A platform may contain some or all of the following features:

- Fire extinguishers
- Fire hose reels
- Fire hydrant systems
- Automatic sprinkler systems

Most fires start as a small fire and may be extinguished if the correct type and amount of extinguishing agent is applied whilst the fire is small and controllable (Service 2012). The standard fire extinguisher types currently available are listed in Table 14.1.

Fire extinguisher locations must be clearly identified. Extinguishers are colour coded according to the extinguishing agent.

TABLE 14.1 Fire extinguishers currently available

EXTINGUISHING AGENT	PRINCIPLE USE
Water	Wood and paper fires; not electrical
Foam	Flammable liquid fires; not electrical
Carbon dioxide	Electrical fires
Dry chemical	Flammable liquids and electrical fires
Wet chemical	Fat fires; not electrical
Special purpose	Various (e.g. metal fires)

There is also a need for a firefighting plan. Fire safety certification is an essential part of the offshore installation. This is determined by the flag state of the structure. An up-to-date fire protection plan, containing a complete inventory and maintenance details of all fire protection components, including firestops, fireproofing, fire sprinklers, fire detectors, fire alarm systems, fire extinguishers, and so forth, are typical requirements for demonstration of compliance with applicable laws and regulations.

All the personnel on the offshore platform should know the emergency procedures in the case of fire. The fire procedure must be reported in the Contingency Plan and must be explained to the personnel. Written procedures are required on the installation to provide the personnel with a guideline in the case of fire or an explosion. It should also be mentioned here that no two fires are ever the same.

An additional aid should come from standby vessels. These vessels have equipment such as a fire monitor in order to tackle external fires and aid the evacuation process. The seamen that occupy these vessels are trained in advanced fire safety (Nutec 2008; Profire 2014).

These vessels should be placed in a safe location. Natural draft components should be equipped with spark and flame arrestors to prevent spark emission.

14.2 EMERGENCY SHUTDOWN SYSTEM AND DEPRESSURISATION

A depressurisation system should be installed to reduce the pressure in the platform, reduce the leak rate, inventory and leak duration and thereby the ignition probability and to reduce the potential for rupture and escalation in a fire by reducing the pressure in the process segment.

The emergency shutdown system (ESD) shall limit the amount of hydrocarbons leaking, thus affecting the extent and duration of fire. ESD is designed to minimize the consequences of emergency situations, related to typically uncontrolled flooding, escape of hydrocarbons or outbreak of fire in hydrocarbon carrying areas or areas which may otherwise be hazardous (Lloyds 2014). The ESD system should be initiated automatically by confirmed gas or fire detection. Depressurisation is either initiated manually or automatically upon confirmed fire detection and normally manually on confirmed gas detection, depending on the statutory and operator requirements for the actual offshore unit. It is recommended to base the design on automatic depressurisation.

Ensuing actions following ESD situations are to alert personnel through the public address system and to provide automatic start-up of platform protection equipment such as fire pumps (Lloyds 2014).

Once an ESD sequence has been initiated and completed, the platform will be in as safe a condition as possible and essential systems will be available so that personnel can tackle the emergency adequately.

The platform ESD system normally interfaces with the following equipment in order to bring it to a safe and steady condition swiftly and effectively:

- Fire and gas system
- Instrument marshalling cabinet
- Platform control system
- Subsea control system
- Computer-aided process operations

14.3 LAYOUT AND VENTILATION

When gas leaks, it will mix with air due to turbulent jet mixing, buoyancy and turbulent interaction with air supplied by ventilation. Molecular diffusion is so slow that it can be discounted. If escaping gas has high velocity, it will develop a cone-shaped momentum jet. This will degenerate into a buoyancy plume. Transition occurs at a point remote from release equal to

$$F = \frac{v_o^2 \rho_o}{\Delta \rho_o g d_o} \tag{14.1}$$

where

F = Froude or Richardson number (i.e. the ratio of inertia forces to buoyancy forces)

$\Delta \rho_o$ = initial density difference between gas and atmosphere

d_o = jet diameter

v_o = initial jet velocity

Experimentation has shown that for methane turbulent momentum jets the LEL (lower explosion limit) will be reached at $\approx$125 orifice diameter along the stream axis.

Theory assumes an infinitely large enclosure. But, reality is that gas will rise to the ceiling where it will spread out to form a layer. As the layer builds up it will no longer be air that is entrained but mixture. Thus, distance travelled before dilution to a given concentration will be increased.

A layer of uniform concentration is formed between the point of leakage and ceiling. The concentration in this layer is dependent on the balance between rate of gas release and rate of supply of fresh air by ventilation (Harris 1983).

14.3.1 Ventilation Regime

Calculate the equivalence ratio, φ, to determine the type of ventilation regime. The equivalence ratio is calculated as

$$\varphi = r \frac{\dot{m}}{\dot{m}_a} \tag{14.2}$$

where
  $\dot{m}$ = mass burning rate of fuel (= mass release rate for gas jet fires)
  $\dot{m}_a$ = the mass rate of entrained air
  r = the mass ratio of air to fuel for stoichiometric burning (r = 17.2 for methane and roughly 15 for heavier hydrocarbons)

Thus, if $\varphi < 1$, then the fire is fuel controlled, and if $\varphi > 1$ the fire is under-ventilated (or 'ventilation controlled').

If $\varphi > 1$ then the fire will not be able to entrain enough air for complete combustion inside the compartment. This is likely to result in increased levels of incomplete combustion products such as CO, increased levels of smoke (soot) and increased flame temperatures, particularly in regions close to the flame/air interface or close to the ceiling of a compartment where hot combustion products may be trapped and recirculated (Lloyds 2014).

14.4 DRAINAGE SYSTEMS

The purpose of the open drain system is to collect liquid leaks and spills and to provide proper disposal to a containment tank. The open drain system in a module should also be able to handle the design fire water capacity in the area (Lloyds 2014).

14.5 FIRE DETECTORS

Fire detectors are installed to provide the necessary alarms upon detection of flame, heat or smoke.

Fire line automatic system can be used to detect fire in open spaces and to activate alarm and/or firefighting equipment automatically. It can be strung around the outside of the entire structure to form a loop.

14.6 GAS DETECTION

Gas detection systems gives an alarm and start protective action when gases exist in dangerous concentrations in various areas of the installation.

Detectors should be located at vantage points to sense the level of gas concentration and transmit a corresponding signal to dedicated control modules located on the fire and gas panels in the control rooms (Lloyds 2014).

14.7 IGNITION PREVENTION DEVICES

Equip components with spark and flame arrestors to prevent spark emission. Reference should be made to API RP 14C for safety systems for fired components. Protection from ignition by electrical sources should be provided by designing and installing electrical equipment using the area classifications as designated by a standard such as IPS-E-SF-120.

14.8 FIRE PUMPS

Provide at least two independently driven power pumps; each pump should be arranged to draw water directly from the sea and discharge into a fixed fire main.

Based on the requirements of standards, booster pumps and storage tanks may be provided for units with high suction lifts. One pump should at least be readily available and dedicated for firefighting duties. The pumps, source of power and sea suctions should be arranged such that fire at any location will not put the pumps out of action. The capacity of the installed pump should be greater than 180 m^3/h.

14.9 FIRE HYDRANTS

A fire hydrant, also called a fireplug or simply a plug, is a connection point by which firefighters can tap into a water supply. It is a component of active fire protection.

Offshore fire hydrants should be positioned at the main access ways at the edge of the platform for each module. The quantity and location of the fire hydrants are to be such that at least two jets of water, not coming from the same hydrant, one of which is to be from a single length of fire hose, shall reach any part of the unit normally accessible to those on board while the unit is being navigated or is engaged in drilling operations. A hose is to be provided for every hydrant.

Fire hoses are to be of approved standard and be sufficient in length to project a jet of water to any of the spaces in which they are required to be used. Their length in general is not to exceed 25 m. Every fire hose is to be provided with a dual purpose jet/spray nozzle and the necessary couplings. Fire hoses together with any necessary fittings and tools are to be kept ready for use in conspicuous positions near the water service hydrants or connections (refer to IPS-E-SF-120).

14.10 PERMIT-TO-WORK SYSTEM

A permit-to-work system is a formal written system used to control certain types of work that are possibly hazardous. The permit to work system

a. Specifies the work to be done and the equipment to be used
b. Specifies the precautions to be taken when performing the task
c. Gives permission for work to start

d. Provides a check to ensure that all safety considerations have been taken into account, including the validity of permits and certificates and compliance to the company's policies and procedures

e. Provides a checking mechanism that all work has been completed to the personnel satisfaction

Permits to work form an important part of safe systems of work for many maintenance activities. There is a need to train personnel on how to prepare and manage the permit-to-work system.

A health and safety survey showed that one-third of all accidents in the chemical industry were maintenance related, the largest single cause being a lack of, or deficiency in, the permit-to-work systems.

A permit to work is required for all work performed involving hot work and/or is in confined spaces. Any employee responsible for the work permit should have been trained and has satisfactory knowledge of the hazards at a work site to be able to specify a system to eliminate, as far as reasonably practicable, the risks in a particular job (HSE 2005).

The issue of a permit does not, by itself, make a job safe. That can only be achieved by those preparing for the work, those supervising the work and those carrying it out. In addition to the permit-to-work system, other precautions may need to be taken; for example, process/electrical isolation, or access barriers, and these will need to be identified in task risk assessments before any work is undertaken. The permit-to-work system should ensure that authorised and competent people have thought about foreseeable risks and that such risks are avoided by using suitable precautions (HSE 2005).

For example, in the permit-to-work system in place at the time of the Piper Alpha disaster, there was no cross-referencing when the work carried out under one permit affected the work under another. Reliance was placed on the memory of the designated authority. Therefore, a permit-to-work system will be more effective if site management and other personnel have been consulted. Imposing systems without consultation can lead to procedures that do not reflect the needs of maintenance staff.

Those carrying out the job should think about and understand what they are doing to carry out their work safely and take the necessary precautions for which they have been trained and made responsible.

Copies of a permit to work should be clearly displayed

a. At the work site, or in a recognised location near the work site. (If this is not practicable, for example, when a job is carried out in a number of locations, then the permit should be kept on the performing authority.)
b. In the central or main control or permit coordination room, with additional copies at any local control rooms.

c. In addition, a copy of the permit should be kept with the issuing authority or with the area authority if that person is not located at the work site or control room (HSE 2005).

14.11 AUTOMATIC FIRE SUPPRESSION

Automatic fire suppression systems control and extinguish fires without human intervention. Examples of automatic fire suppression systems include fire sprinkler system, gaseous fire suppression and condensed aerosol fire suppression. When fires are extinguished in the early stages, loss of life is minimal since 93% of all fire-related deaths occur once the fire has progressed beyond the early stages.

In general, however, automatic fire suppression systems fall into two categories: engineered and pre-engineered systems (Wikipedia 2017).

14.11.1 Automatic Fire Sprinklers

A fire sprinkler system is an active fire protection method, consisting of a water supply system, providing adequate pressure and flowrate to a water distribution piping system, onto which fire sprinklers are connected.

Automatic sprinklers can be installed in the fixed offshore platform to suppress automatically small fires on, or soon after, an ignition/explosion or to contain fires until the arrival of the fire service. A life safety sprinkler system should be installed in accordance with the appropriate standards.

The main cause of fatalities in fire is smoke, and most deaths occur long before there is any significant risk of structural collapse. Structural damage is normally of secondary importance. By suppressing fire and smoke, sprinklers are an extremely effective means of enhancing life safety and reducing financial losses.

A fire sprinkler system is an active fire protection method. It consists of a water supply system providing adequate pressure and flow rate to a water distribution piping system, onto which fire sprinklers are connected. Sprinkler systems provide early fire control or extinguishment, helping to mitigate the hazards for occupants and firefighters alike.

The three types of sprinkler systems are

a. Standpipe system
b. Wet standpipe system
c. Dry standpipe system

The automatic fire sprinkler should be capable of indicating the presence of a fire in all accommodation spaces and service spaces.

Where such a system is fitted, it is to be of the wet pipe type, but small exposed sections may be of the dry pipe type where this is shown to be a necessary precaution. Any part of the system which is subjected to freezing temperatures in service is to be suitably protected against freezing. It is to be kept charged at the necessary pressure and have provision for a continuous supply of water.

References

Abdollahzadeh, G. R. and Nemati, M. n.d. Risk Assessment of Structures Subjected to Blast. Accessed August 28, 2017. www.researchgate.net/publication/274509776_Risk_Assessment_of_Structures_Subjected_to_Blast.

AISC. 1991. *Specification for Structural Steel Buildings – Allowable Stress Design and Plastic Design.* American Institute of Steel Construction (AISC).

AISC. 2011. *Steel Construction Manual,* 14th edition. ISBN 1-56424-060-6. United States.

AkzoNobel. 2017, June 20. Protective coatings. Retrieved from AkzoNobel: http://www.international-pc.com/products/fire-protection/fire-protection-technical-information.aspx

Ali, R. M. 2007. *Performance-Based Design of Offshore Structures Subjected to Blast Loading.* London: Imperial College.

André, J., Besle, R., Baptista, A. 2015. New indices of structural robustness and structural fragility. *Structural Engineering & Mechanics,* 56. doi: 10.12989/sem.2015.56.6.1063.

API. (2000). *API Recommended Practice (2A-WSD).* USA: American Petroleum Institute.

Arnold. 2017. Major offshore accidents of the 20th and 21st century. Retrieved October 13, 2017, from http://www.oilrigexplosionattorneys.com/Oil-Rig-Explosions/History-of-Offshore-Accidents.aspx

ASCE 1997. *Design of Blast Resistant Buildings in Petrochemical Facilities.* New York: American Society of Civil Engineers.

ASFP. 2015. Types of fire protection materials. Retrieved October 14, 2017, from http://www.asfp.org.uk/Technical%20Services/types_of_fire_protection.php

Baker, J.W., Schubert, M., Faber, M.H. 2007. On the assessment of robustness.

Bea, R. G. 1991. *Design and Characterize the Offshore Fire Problem. Improved Means of Offshore Platform Fire Resistance.* Berkeley: University of California.

Biggs, J.M. 1964. *Introduction to Structural Dynamics.* New York: McGraw-Hill.

Bowles, J. E. 1958. *Foundation Analysis and Design.* New York: McGraw-Hill.

Brode, H.L. 1955. Numerical solution of spherical blast waves. *Journal of Applied Physics.*

Chakrabarti, S. 2005. Handbook of offshore engineering. In K. Demir, *Fixed Offshore Platform Design* (pp. 279–401). USA: Elsevier.

Chopra, A.K. 2007. *Dynamic of Structures: Theory and Applications to Earthquake Engineering,* 3rd edition. Upper Saddle River, New Jersey: Pearson/Prentice Hall.

Corr, R.B., Tam, V.H.Y. 1998. "Gas explosion generated drag loads in offshore installations." *Journal of Loss Prevention in the Process Industries,* 43–48.

DnV. 2014. *Lifting Operations (VMO Standard – Parts 2–5).* Norway: Det norske Veritas.

DnV. June 2016. DNV-OS-H202: Sea transport operations (VMO Standard – Part 2-2). Det Net Veritas.

Eknes, M.L. and Moan, T. 1994. Escalation of gas explosion event offshore. *Offshore Structural Design, Hazard, Safety and Engineering, Conference Proceedings London.* ERA Report No 94-0730, 3.1.1–3.1.16.

EN1990:2002. 2002. *Basis of Structural Design.* European Commitee for Standardization (CEN), Eurocode.

Engelhardt, M. D. 2016. *Basic Concept in Ductile Detailing of Steel Structures.* Austin: University of Texas.

ESDEP. n.d. ESDEP 15A. Retrieved July 5, 2017, from http://fgg-web.fgg.uni-lj.si/%7E/pmoze/esdep/master/wg15a/l0100.htm

ESDEP. n.d. ESDEP 4B.3. Retrieved July 9, 2017, from http://fgg-web.fgg.uni-lj.si/%7E/pmoze/esdep/master/wg04b/t0300.htm

ESDEP. n.d. ESDEP 4B.4. Retrieved July 10, 2017, from http://fgg-web.fgg.uni-lj.si/%7E/pmoze/esdep/master/wg04b/t0400.htm

ESDEP. n.d. ESDEP WG 8. Retrieved June 10, 2017, from http://fgg-web.fgg.uni-lj.si/%7E/pmoze/esdep/master/wg08/l0420.htm

ESDEP. n.d. ESDEP WG 15A. Retrieved July 10, 2017, from http://fgg-web.fgg.uni-lj.si/%7E/pmoze/ESDEP/master/wg15a/l0800.htm

ESDEP. n.d. ESDEP WG 4B. Retrieved July 6, 2017, from http://fgg-web.fgg.uni-lj.si/%7E/pmoze/esdep/master/wg04b/t0100.htm

ESDEP. n.d. Punching shear. Retrieved July 10, 2017, from https://www.sharcnet.ca/Software/Ansys/17.0/en-us/help/wb_beam/API_ALL_PUNC_SUC.html#Section5.4.4.1

Eurocode 3: Design of steel structures – Part 1-1: General rules and rules for buildings, BS EN 1993-1-1: 2005.

Farid Alfawakhiri, Longinow Anatol. 2003. "Blast resistant design with structural steel–common questions answered." 61–66. Modern steel construction.

GPB. 2010. *Guidelines for Marine Lifting Operation – 0027/ND.* London: Noble Denton Group Limited.

Harleman, D. R. F. 2010. *Dynamic Analysis of Offshore Structures.* Massachusetts: Elsevier, 482–499.

Harris, R.J. 1983. *The Investigation and Control of Gas Explosions in Buildings and Heating Plant.* University of Michigan: E. & F.N. Spon in association with the British Gas Corp.

HSE. 2005. *Guidance on Permit-to-Work System.* London: Health and Safety Executive.

IGN – Interim Guidance Notes for the Design and Protection of Topside Structures against Explosion and Fire. Steel Construction Institute, 1992.

ISO19902. 2007. Strength of tubular members. In *ISO19902, Petroleum and Natural Gas Industries – Fixed Steel Offshore Structures* (pp. 100–112). International Standard Organization.

Izadifard, R. A., and Maheri, M. 2010. Ductility effects on the behaviour of steel structures under blast loading. *Iranian Journal of Science & Technology*, 49–62.

Izzuddin, B.A. and Lloyd Smith, D. 1997. Response of offshore structures to explosion loading. *International Journal of Offshore and Polar Engineering*, 212.

James, M. A. 1988. Analytical methods for determining fire resistance of steel members. In *SFPE Handbook of Fire Engineering (Chapter 6)*, 1st edition. Quincy, MA: Society of Fire Protection Engineers.

Kövecsi, T. 2014. *Membrane Action of Slabs in Framed Structures in Case of Accidental Column Loss*. Romania: Politehnica University Timişoara.

Kharade, A.S. and Kapadiya, S.V. 2014. "Offshore engineering: an overview of types and loadings on structures." *International Journal of Structural and Civil Engineering Research*, 1–13.

Ki-Yeob Kang, Kwang-Ho Choi, JaeWoong Choi, YongHee Ryu, Jae-Myung Lee 2016. Dynamic response of structural models according to characteristics of gas explosion on topside platform. *Ocean Engineering*, 174–190. doi: 10.1016/j.oceaneng.2015.12.043.

Lloyd's. August 2013. *Code for Lifting Applications in a Marine Environment*. London: Lloyd's Register Group Limited.

Llyods. 2014. *Guidance Notes for Risk Based Analysis: Fire Loads and Protection*. London: Lloyd's Register.

Mills, C. A. 1987. The design of concrete structure to resist explosions and weapon effects. *Proceedings of the First International Conference on Concrete Hazard Protections* (pp. 61–73). Edinburgh, UK.

Moore, D. B. and Wald, F. 2003. *Design of Structural Connections to Eurocode 3*. Watford: Leonardo Da Vinci, 1–136.

Nallayarasu, S. 2013. *Offshore Structures: Design and Analysis*. Madras: Indian Institute of Technology.

Ngo, T., Mendis, P., Gupta, A., and Ramsay, J. 2007. "Blast loading and blast effects on structures – An overview." *EJSE Special Issue: Loading on Structures*, 79–80.

NORSOK. 2008. *Integrity of Offshore Structures*. Norway: Standards Norway.

Nutec, F. 2008. *Basic Safety Offshore*. Netherlands: Falck Nutec.

Paik, J.K. 2011. *Quantitative Assessment of Hydrocarbon Explosion and Fire Risks in Offshore Installations*. Elsevier, pp. 73–96.

Profire. 2014. Section factors. Retrieved June 20, 2017, from W/D, A/P, M/D Calculation Method: http://www.profire.com.tr/eng/222-page-wd-ap-md-calculation-method.aspx

Qiao, A., Zhang, S. 2010. Advanced CFD modeling on vapour dispersion and vapour cloud explosion. *Journal of Loss Prevention in the Process Industries*, 23, 843–848.

Raabe, D. n.d. Dramatic failure of materials in drilling platforms. Retrieved August 12, 2017, from http://www.dierk-raabe.com/dramatic-material-failure/oil-drilling-platforms/

Risanto, P. A. 2015, September 6. Linkedin Slideshare. Retrieved July 31, 2017, from https://www.slideshare.net/PuputAryanto/introduction-to-oil-gas-health-safety-environment

Robertson, A.F., and Ryan, J.V. 1995. Proposed criteria for defining load failure of beams, floors and roof constructions during fire tests, *Journal of Research, National Bureau of Standards*, 63C, 121.

Service, S. A. 2012, May 24. Fire fighting systems and equipment in buildings. Retrieved September 7, 2017, from http://www.mfs.sa.gov.au/site/community_safety/commercial/building_fire_safety/fire_fighting_systems_and_equipment_in_buildings.jsp

SJR. 1994. *Subsea Structure Design*. London: JP Kenny Group.

Starossek, U., Haberland, M. 2008. *Approaches to Measures of Structural Robustness*. Seoul, Korea.

Stolz, A., Klomfass, A., Millon, O. 2016. "A large blast simulator for the experimental investigation of explosively loaded building components." *Chemical Engineering Transactions*, 151–156. doi: 10.3303/CET1648026.

Szarka, I. 2015. *Structural Integrity Management Ensuring Robustness and Barriers.* Stavanger: University of Stavanger.

Tam, V.H.Y. and Simmonds, S.A. 1990. Modelling of missile energy from gas explosions offshore. *International Conference on the Management and Engineering of Fire Safety and Loss Prevention Onshore and Offshore.*

TM5. 1990. *Design of Structures to Resist the Effects of Accidental Explosions.* Washington, DC: Department of Army – TM5-1300.

TMR. 2009. *Resistance to Accidental and Catastrophic Fires: General Principles.* Norway: Norwegian University of Science and Technology.

US Army. 30 October 1992. Solution of bearing capacity. In *Engineering and Design: Bearing Capacity of Soils* (p. 39). Washington, DC: Department of the Army.

Vasilis Karlos, G.S. 2013. *Calculation of Blast Loads for Application to Structural Components.* Italy: European Commission – Joint Research Centre.

Visser, W. June 11–16, 1995. Simplified dynamic assessment for fixed offshore structures under extreme waves. *Proceedings of the Fifth (1995) International Offshore and Polar Engineering Conference* (p. 1). The Hague, The Netherlands: International Society of Offshore and Polar Engineers.

Wikipedia.contributor. n.d. Automatic fire suppression. Retrieved September 7, 2017, from Wikipedia: https://en.wikipedia.org/w/index.php?title=Automatic_fire_suppression&oldid=790368428

Wikipedia.contributor. n.d. Piper Alpha. Retrieved September 6, 2017, from Wikipedia: https://en.wikipedia.org/w/index.php?title=Piper_Alpha&oldid=799012771

Wikipedia.contributor. n.d. Structural dynamics. Retrieved September 23, 2017, from Wikipedia: https://en.wikipedia.org/w/index.php?title=Structural_dynamics&oldid=779299356

Index